T0317486

Complexities 1

Systems of Systems Complexity Set

coordinated by
Jean-Pierre Briffaut

Volume 5

Complexities 1

*Various Approaches in the Field of
Techno-scientific Knowledge*

Edited by

Jean-Pierre Briffaut

Foreword by

Philippe Kourilsky

WILEY

First published 2023 in Great Britain and the United States by ISTE Ltd and John Wiley & Sons, Inc.

Apart from any fair dealing for the purposes of research or private study, or criticism or review, as permitted under the Copyright, Designs and Patents Act 1988, this publication may only be reproduced, stored or transmitted, in any form or by any means, with the prior permission in writing of the publishers, or in the case of reprographic reproduction in accordance with the terms and licenses issued by the CLA. Enquiries concerning reproduction outside these terms should be sent to the publishers at the undermentioned address:

ISTE Ltd
27-37 St George's Road
London SW19 4EU
UK

www.iste.co.uk

John Wiley & Sons, Inc.
111 River Street
Hoboken, NJ 07030
USA

www.wiley.com

© ISTE Ltd 2023
The rights of Jean-Pierre Briffaut to be identified as the author of this work have been asserted by him in accordance with the Copyright, Designs and Patents Act 1988.

Any opinions, findings, and conclusions or recommendations expressed in this material are those of the author(s), contributor(s) or editor(s) and do not necessarily reflect the views of ISTE Group.

Library of Congress Control Number: 2023941801

British Library Cataloguing-in-Publication Data
A CIP record for this book is available from the British Library
ISBN 978-1-78630-875-7

Contents

Chapter 2. Complexity and Biology: When Historical Perspectives Intersect with Epistemological Analyses 51
Céline CHERICI

Chapter 3. Two Complexities: Information and Structure Content 87
Jean-Paul DELAHAYE

Chapter 4. Leveraging Complexity in Oncology – A Data Narrative

Xosé M. FERNÁNDEZ

Chapter 5. Complexity or Complexities of Information: The Dimensions of Complexity

Jacques PRINTZ

Foreword

Sharing Complexity:
An Acclaim for Complex Thinking

F.1. Omnipresent complexity

Complexity is everywhere, all the time. We are all confronted with it and are constantly faced with complex situations and problems to which we find solutions and adapt our behavior. Curiously, we often negotiate these unconsciously more than rationally. Only a fraction of the complex decisions we make reach our consciousness. This is a property of our brain and our organic form: our immune system (a highly complex system) operates day and night outside our conscious awareness. And that is a good thing: it saves us a lot of energy.

However, beyond this biological consideration, is it not strange that, in the face of complexity, we have so neglected our conscious mental activity? Complexity should be part of our standard intellectual arsenal, one of the major notions we should all learn at school, and one we should all share. But it is not.

Why is this? First, we often find it more convenient to ignore it. This is understandable: complexity is tiring, and taking it seriously takes time and effort. However, we also find it more appropriate and rational to ignore or reduce it. Let us not forget that it is on the basis of the deliberate reduction of complexity that much contemporary science has been built, notably classical physics and molecular biology. Today, however, this reductionism is far from being as exclusive as it was: complexity has taken its place in the "hard" sciences.

F.2. The science of complexity

The science of complexity has been born. It is a recent development, since the turning point was reached in the second half and especially toward the end of the 20th century. It covers an ever-increasing number of fields, as shown by the dazzling development of informatics, algorithms and computers combined with the increasing complexity of human-made artifacts (cars, airplanes, etc.). Artificial intelligence is a striking manifestation of this. Engineering has been turned upside down. So have the natural sciences. Increased data collection, storage, and analysis capacities have given climatology a whole new dimension. The same is true of biology, since it passed the symbolic and practical milestone of human genome sequencing in the early 2000s.

The movement does not stop at the "hard" experimental sciences. It is now reaching the social sciences. On our finite planet, the complexity of human society is increasing very rapidly due to a demographic growth whose consequences are often underestimated (there are 10 times more of us today than in 1789), as well as the multiplication of physical and virtual links between individuals and social groups (over 4 billion people are connected to the Internet, and even more have a cell phone).

The science of complexity is not yet a well-identified discipline, and this raises the question of transversality and trans-disciplinarity. As our dialogue with Jacques Printz[1] shows, the exchange is as productive and fascinating as it is difficult: the different disciplines have developed their own methods and languages, making mutual understanding as essential as it is delicate.

F.3. The role of chance

Complexity calls for certain modes of reasoning. One of the most important of these concerns the distinction between what is complicated and what is complex. In our view, one of the things that sets them apart is that the complex makes room for chance. Chance may be "happy" or "unhappy", but it is always present. This is the basis of the theory of the evolution of species, which rests on the notion that mutations in DNA are unpredictable (but not necessarily equiprobable). An airliner is complex, not just complicated, because from time-to-time unforeseen accidents occur, often unpredictable and at random.

---

1 See the afterword in Printz, J. (2023). *Organization and Pedagogy of Complexity: Systemic Case Studies and Prospects*. ISTE Ltd, London, and Wiley, New York.

The possible origins, for an aircraft as for a human being, are either external (from the environment; for example, a storm for an aircraft or a virus for a human being) or internal (a fire for an aircraft or cancer for a human being). In both cases, a large number of malfunctions are resolved through mechanisms that detect and correct them before they become problematic for the system as a whole. This property, which complex systems of all kinds possess to varying degrees, is known as "robustness". Robustness is a major concept because it describes a system's ability to function "correctly" despite external or internal hazards.

The more complex a system, the more components and links it comprises, the greater the "chances" of error and the more monitoring and control devices are required. Errors are inevitable. They can be reduced, but not totally eliminated. Perfection would mean multiplying controls to such an extent that their abundance would eventually paralyze the system. The human organism spends a great deal of resources defending itself against diseases such as microbial infections and cancer. It succeeds in the vast majority of cases: only those that have escaped a multitude of internal controls become manifest. Although it cannot control them all, our organic form should be considered "robust". Its robustness is our insurance against life. It defends us against life's hazards, enabling us not to live, but to survive. Without it, our lives would be short-lived.

F.4. An acclaim for complex thinking

Complex thinking is not only the work of philosophers: it is also, and perhaps above all, a by-product of the so-called hard sciences, as we have shown and stated in our last two books[2]. As complexity invades our daily lives, we need to learn to share its management and to "think complex".

The ubiquity of complexity is nothing new. What is new is its growing involvement in social and technical fields, as tools emerge to better describe and manage it. New behaviors, new lifestyles and new professions are emerging.

That is why it is so important to make complex thinking our own. It is singular in more ways than one. It is more than a purely intellectual process. It must accommodate a certain amount of uncertainty inherent to complex systems and requires intellectual navigation between the whole and its parts. It encourages tolerance rather than dogmatism. It requires discussion. It requires strong ethics, because it is open to all manipulations that play on uncertainty and make it permeable to fake news and intellectual dishonesty.

---

2 Kourilsky, P. (2014). *Le Jeu du hasard et de la complexité*. Odile Jacob, Paris; Kourilsky, P. (2019). *De la science et de la démocratie*. Odile Jacob, Paris.

F.5. Complexity and democracy

The growing complexity of the world has major political impacts. It is too often overlooked that complexity is vital to democracy. By giving many the impression that democratic regimes can no longer manage it satisfactorily, complexity has become a real citizenship issue. Issues of democracy are becoming increasingly pressing to the point where we have to ask whether it is robust enough (in the sense of the robustness of complex systems) to withstand the blows of nationalism, populism and authoritarianism, and this at a critical time when we are faced with a globalization that perpetuates and produces enormous inequalities, as well as a very serious environmental crisis.

Mastering complexity by practicing complex thinking is therefore doubly essential for our mutual future: we need to apprehend and manage increasingly complex situations, which call for adapted modes of analysis and action, but also socially difficult and potentially conflictual arbitrations. We must also succeed in doing so within the democratic framework to which we are committed.

That is why, as a citizen as much as a scientist, we offer a vibrant acclaim for complex thinking.

Philippe KOURILSKY
Member of the Académie des Sciences
and Honorary Professor at the Collège de France

Preface

Why put in the title of this book "complexities" in the plural?

The word "complexity" is used in a wide variety of contexts to account for situations in very different realms of knowledge.

There are many ways to quantify our challenges:

– design an object with many features;

– perform a task;

– understand the uses of a system with correlated functions;

– understand the tree structure of a document search with hypertext links;

– decipher a message;

– become aware of an operational drift situation;

– understand the motivations of the actors in a collaborative context;

– react in a crisis situation;

– counter the emergence of complexity, that is, facilitate the transition to simplification in the realm of thought and discourse.

In each case, approaches to attempt to control the situation may involve very different methods and techniques.

It is the ambition of this book to try to show, without aiming at a complete panorama, these varieties of approaches and behaviors vis-à-vis the obstacles that constitute complexity in the understanding and realization of human courses of action. Sometimes, for the cultural sciences in particular, these obstacles can lead to wealth insofar as the analysis leads to cross-fertilization interactions between disciplines that do not usually collaborate.

In general, it is worth noting that the specialized literature shows a clear reciprocity between the concepts of "system" and "model". It is through astronomy that the notion of system, first understood as a material organisation in ancient Greece, has been used in the realm of knowledge. It is the Galilean revolution that brought this term into the semantic field of abstract theories (*The Dialogue Concerning the Two Chief World Systems* by Galileo (1632)).

As for the word "model", it comes from the Latin *modulus*, diminutive of *modus*, "measure". It was initially a term of architecture used to establish the relationships of proportion between the parts of an architectural work. The word "*modulus*" led in the Middle Ages and the Renaissance first to the word "mould", then in the 16th century to the word "model" through the intermediary of the Italian "*modello*".

The word "model" presents a congenital ambiguity, sometimes meaning the original, the ideal to achieve, and sometimes the copy, the simple realization or interpretation of an existing entity. Widely used in experimental sciences from the 19th century onwards, it appears as an instrument of intelligibility "the function of which is a function of delegation. The model is an intermediary to which we delegate the knowledge function, more precisely the still-puzzling enigma, in the presence of a field of study the access of which, for various reason is difficult to us"[1].

The themes covered in the different chapters make this clear explicitly or implicitly; the concepts of "system" and "model" have a heuristic function in approaching complexity in many fields of knowledge.

The contributions are divided into two parts. The first includes contributions with technical and scientific orientations. The second deals with subject matters that are more related to the human and social sciences.

Jean-Pierre BRIFFAUT

July 2023

---

[1] Bachelard, S. (1979). Some historical aspects of the notions of model and the justification of models. In *Proceedings of the Colloquium Elaborating and Justifying Models*. Maloine-Doin, Paris.

The Complexity of Cybersecurity

1.1. Formal approach to the complexity of cybersecurity

1.1.1. *Cybersecurity and theoretical computing*

Born with computers, digital insecurity has accompanied their development for more than 60 years with each algorithmic or hardware innovation. The first computer viruses appeared with the first programming languages, just as the first vulnerabilities weakened the first networks. Thinking about the security of a computer system requires a systemic approach that takes into account the strong heterogeneity of the components of the system and the complexity of the interactions that animate it. Any digital system must therefore be considered as a combination of three types of interacting components: "hardware" components, "software" components (software, data) and "human biological" components (user). It is important to include the human user as an integral component of the system whose security is to be evaluated or certified, as many successful cyberattacks exploit the human factor as a weak link in the security chain.

A cyberattack can be represented by a quadruplet $A = (C_A, O_A, S_A, H_A)$ in which C_A designates the target(s) of the attack, O_A designates the attacker or the group of attackers carrying out the attack, S_A designates all the malware implemented by the attacker during the attack and H_A designates all the hardware components used by the attacker.

The target $C_A = (O_C, S_C, H_C)$ is a triplet made up of the O_C set of the operators or human users impacted, the S_C set of the targeted programs or software of the

Chapter written by Thierry BERTHIER and Thomas ANGLADE.

information system and the H_C set of the targeted hardware components. Once executed, the attack A applied to a C_A target produces a pair (gain for the attacker, loss for the target) represented by: $Attack < C_A > = (Gain(O_A), Loss(C_A))$, which represents the impact of A.

With this preliminary formalism in place, we are going to show that absolute digital security does not exist. For this, we must use the Turing machine concept.

A Turing machine is an abstract model of the functioning of a computer and its memory created by Alan Turing in order to give a precise definition to the concept of an algorithm.

A Turing machine consists of three parts as follows:

1) A storage unit or ribbon of infinite length divided into cells. Each cell contains one symbol from a finite alphabet.

2) A read-write head that moves one cell at a time to the right or left along the ribbon. The reading head reads or writes symbols.

3) A control function driving the read-write head. The control function in turn consists of a status register storing the current state of the machine and an action table – the program. The number of possible states is finite. The action table tells the machine which action to perform according to the contents of the status register and the current cell.

DEFINITION (TURING MACHINE).– A Turing machine M is defined by:

- a set of $n+1$ states $S_M = \{s_0, s_1,..., s_n\}$ with $n \in N$;

- a set of $m+1$ symbols $I_M = \{i_0, i_1,..., i_m\}$ with $m \in N$;

- a set $d = \{-1,0,+1\}$ describing the possible movements for the reading head;

- an output function $O_M: S_M \times I_M \rightarrow I_M$;

- a transition function $N_M: S_M \times I_M \rightarrow S_M$;

- a motion function $D_M: S_M \times I_M \rightarrow d$.

The machine M is then referred to as the quintuplet (S_M, I_M, O_M, N_M, D_M).

The set of Turing machines will be denoted as $\mathcal{M}$.

We therefore have, in accordance with the three parts detailed above, a finite number of possible states S_M, a finite alphabet I_M, the definition of the possible movements of the reading head d and the control functions O_M, N_M and D_M.

The notion of the history $H_M(t)$ of a Turing machine M is then defined from three time functions, making it possible to give, for each time index:

1) the state of the machine at the end of the action according to the step index, called the temporal function of state $E_M : N \rightarrow S_M$;

2) the cell content as a function of step index and cell number, called the band time function $B_M : N \times N \rightarrow I_M$;

3) the cell number after the displacement of the read head according to the step index given by the cell time function $C_M : N \rightarrow N$.

Thus, the history $H_M(t)$ at a given instant t of a Turing machine M is defined by:

$H_M(t) = (E_M(t), B_M(t), C_M(t))$; the initial state of M is given by $H_M(0)$.

To these first definitions are added the structure TP_M describing a Turing machine (M) program and the non-empty set of Turing programs TS.

1.1.1.1. *The halting problem*

Given an arbitrary program P, we will show that there is no systematic way to tell if this program will end on an input or if it will loop and calculate indefinitely on this input without ever stopping or giving a result. This major result obtained by Turing may seem trivial to the first time discoverer, but it is fundamental and forms the basis of other results of great depth. Here it can be demonstrated again at little cost.

We proceed via the absurd and assume that we have a program A that can analyze a program P and an input x in the following way: $A(P, x) = 1$ if P ends when it is given x as input and $A(P, x) = 0$ if P does not end when x is given as input.

Our program A is therefore able to detect for all existing and future programs P in the world whether they stop at an x input or not.

From this program A, we will build a super program B which does something very simple:

By definition of B:

If $A(x, x) = 0$, then $B(x) = 1$;

If $A(x, x) = 1$, then $B(x)$ runs indefinitely without ever stopping and without giving any result.

Building program B from our initial program A is very simple; we write:

```
int B(Program x)
{
if (A( x , x ) == 0) return (1)
else
{ while(true) { } }
}
```

Once this program B has been constructed, we ask ourselves about the value of B (B), that is, the value of B applied to B. What is the value of B(B)?

There are only two possible cases:

Case no. 1: B(B) = 1.

If B(B) = 1, then going back to the definition of B, we see that A(B, B) = 0, which means that B does not stop (it loops forever) when we give it B as input. This is absurd since B(B) = 1.

Case no. 2: B(B) does not stop.

If B(B) does not stop, then A(B,B) = 1 and B(B) ends, which is still absurd.

The two possible cases lead to absurdity, which means that our initial hypothesis is never verified. Our program A does not exist!

1.1.1.2. *Absolute digital security does not exist!*

It is often said that there is no such thing as absolute IT security. More than just a pessimistic slogan, this is a systemic truth based on a mathematical result related to Turing's machine halting problem.

If this absolute security existed, we would have an oracle analyzer M capable of telling without ever being wrong for any program P whether it is safe or not. In functional terms, we would have for any program P: M(P) returns 1 if the program is judged safe by M and M(P) returns 0 if it is judged unsafe by M.

To say that absolute security exists is to say that the magic analyzer exists. It is also saying that it has not yet been found today but that in a while (years or centuries), after laborious algorithmic efforts, it will be discovered. Let us show that this oracle analyzer does not exist and will never exist even in the distant future.

To show that *M* does not exist, we will use the result of the Turing halting problem.

Most of the work has been done in the section dealing with the halting problem. We are going to go through the absurd again. Let us suppose that the security company *AbsoluteSecurity* one day develops a magic analyzer *M* that can detect whether any program *P* is safe or not, and that it can do this for any program *P*.

Let us specially build the following short program with only three lines. In the order of execution of this program, the first line generates a very secret key called "Verysecretkey". The next line calls up any P' program and the third line prints and displays the top secret key on the screen:

```
verysecretkey = newkey( );
P'( );
System.out.println( verysecretkey);
```

Now let us look at how this three-line program works.

If the program P'( ) does not loop indefinitely, it necessarily means that this program is totally vulnerable (and therefore unsafe) since it will display our very secret key to the public as soon as P' has finished its execution, that is, after a finite amount of time.

If, on the contrary, the program P'( ) does not stop and loops forever, the third instruction of the program will never be executed and our top secret key will never be revealed. In this case, our program is very safe since it does not display the key.

Now let us use the magic analyzer from *AbsoluteSecurity* on our three-line program. The magic analyzer works for all P-programs, so especially for this program. It is therefore able to tell whether it is safe or not, that is, to decide whether the program P'( ) in the second line stops or not. However, we have shown (halting problem) that this program does not exist.

In conclusion, the security company *AbsoluteSecurity* is arrogant. It lies about its magic analyzer M, which is not able to analyze all programs. Absolute security does not exist!

In an information system, security threats and incidents can result from a cyberattack or an accident, as shown in Figure 1.1.

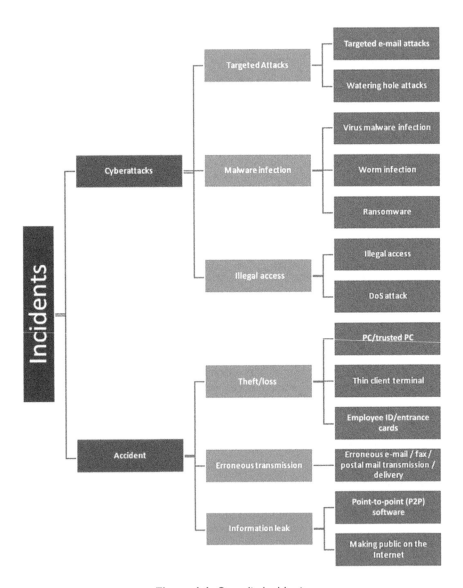

Figure 1.1. *Security incidents*

1.1.2. *Malware and computer virology*

"Malware" is an acronym for malicious software, which refers to any script or binary code that performs some malicious activity. Malware can come in different formats, such as executables, binary shell code, script and firmware. There are several non-equivalent approaches to defining malware:

– *NIST's definition*: a program that is inserted into a system, usually covertly, with the intent of compromising the confidentiality, integrity or availability of the victim's data, applications, or operating system or otherwise annoying or disrupting the victim.

– *TechTarget's definition*: any program or file that is harmful to a computer user. Malicious programs can perform a variety of functions, including stealing, encrypting or deleting sensitive data, altering or hijacking core computing functions and monitoring users' computer activity without their permission.

– *BullGuard's definition*: a computer program designed to infiltrate and damage computers without the user's consent.

– *Kaspersky's definition*: a type of computer program designed to infect a legitimate user's computer and inflict harm on it in multiple ways. Malware can infect computers and devices in several ways and comes in a number of forms, just a few of which include viruses, worms, Trojans, spyware and more.

– *Norton's definition*: malware is software that is specifically designed to gain access to or damage a computer without the knowledge of the owner.

– Or-Meir, Nissim, Elovici and Rokach's definition: "Malware is code running on a computerized system whose presence or behavior the system administrators are unaware of; were the system administrators aware of the code and its behavior, they would not permit it to run" (Or-Meir et al. 2019). Malware compromises the confidentiality, integrity or the availability of the system by exploiting existing vulnerabilities in a system or by creating new ones.

There are several functional criteria to build a malware classification. This can be done simply by the type of malware, or by the behavior of the malware code or by considering the privileges that the malware is running on the system it has just infected (Figure 1.2).

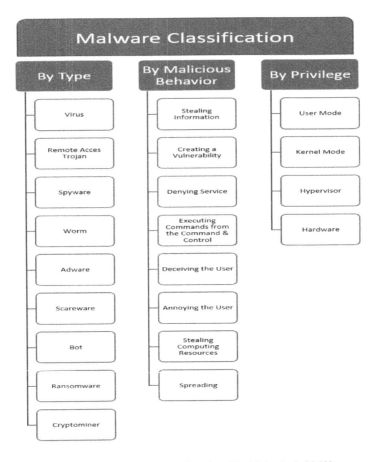

Figure 1.2. *Malware classification (Or-Meir et al. 2019)*

Definitions of the main types of malware are given in the study by Ori Or-Meir et al. (2019). We reproduce the definitions of the most common malware:

– A *virus* injects its malicious code into other files, thus spreading within the host (and potentially to other hosts as well). The term *virus* is often used by the mass media to describe any kind of malware.

– A *remote access Trojan* (RAT, or Trojan horse) is a type of malicious software that pretends to be harmless, like the mythological wooden horse that the Greeks sent to Troy. Most of the time, RATs apply social engineering techniques and hide from the user by disguising themselves as benign software, a useful tool or a game to stay in the background and perform their malicious tasks.

– *Spyware* tracks the user without his/her consent and reports back to the attacker about the user's activities, visited websites, geographic location and so on.

– A *worm* is a program that duplicates itself and spreads through networks. Worms can spread very quickly and disrupt system use by clogging the network. Once the *worm* is detected, it is often easy to patch the system and prevent it from spreading further.

– *Adware* automatically shows advertisements to the user. These ads can be injected into other software or Web pages, and in some cases, *adware* can even replace an existing ad with another ad. Advertising space is sold by the *adware's* author to vendors and companies, generating profits for the author. This type of malware does not usually harm the system, and most of the times the user will never be able to tell that he/she was infected; for this reason, *adware* is also referred to as *grayware*.

– *Scareware* presents an interface that informs users they have been infected with malware. The interface is designed to look like an anti-virus and includes an offer to purchase software to remove the malware. After the victim purchases and installs the software, the *scareware* is removed by the purchased software. While most *scareware* does not harm the infected machine in any way, it deceives the user with false detection messages and can lead the user to believe that an actual anti-virus tool was installed.

– A *bot* (derived from the word robot) is a type of malware that performs actions without the user's consent as part of an army of "zombies". Such actions include visiting websites, spreading the malware to other hosts, and querying a service (such as the domain name system (DNS) or email servers). All the *bots* receive their instructions from a command and control (C&C) server, and the group of infected hosts is referred to as a botnet. Botnets are mainly used for distributed denial-of-service (DDoS) attacks, in which many infected devices (sometimes millions) are used simultaneously to overload a Web service (websites, DNS, cloud services, etc.). By creating a swarm of requests, the target server is flooded and unable to provide service for legitimate users.

– *Ransomware* encrypts the user's files (documents and photos) with a strong form of encryption and demands payment in exchange for the decryption key. Usually, this type of behavior does not prevent use of the computer, but it effectively renders all of the information on the computer inaccessible. Users who have failed to back up their files before the attack and its file encryption are left helpless and forced to choose between paying a high ransom or giving up completely, formatting the entire host and losing everything. To make things worse, most *ransomware* demands payment within a short timeframe (usually a few days). Modern *ransomware* demands payment in Bitcoin or other cryptocurrencies because they allow the attacker to remain relatively anonymous.

– *Cryptominers* use cryptocurrency differently than ransomware. Instead of forcing infected users to pay a ransom to the attacker, the malware uses any available computing power of the victim to mine cryptocurrencies for the attacker. The victims bear the brunt of the increased electricity costs and performance deterioration, while the attacker makes all the profit. This is a fairly new attack that has been on the rise since 2017; reportedly, almost 90% of recent malware attacks are *cryptominers*. This trend is explained by the fact that mining cryptocurrencies yields great profits to attackers while requiring minimal effort.

The relationships between the types of malware and their malicious behavior are given in Table 1.1.

		Malicious Behavior							
	→	Spreading	Denying Services	Creating a Vulnerability	Executing Command from the C&C	Deceiving the User	Stealing information	Annoying the User	Stealing Computing Resources
	Virus	X							
	Worm	X	X						
Type	Ransomware		X						
of	Bot		X	X	X				
Malware	Remote Access Trojan			X	X	X	X		
	Scareware					X			
	Spyware						X		
	Adware							X	
	Cryptominer								X

Table 1.1. *Relationships between types of malware and malicious behavior*

In order to escape the protective shields and analysis systems of the machines they infect, malware often behaves very stealthily once installed on targets. Figure 1.3 shows these stealthy and intrusive behaviors.

We show in the second part how machine learning techniques can effectively detect certain malware.

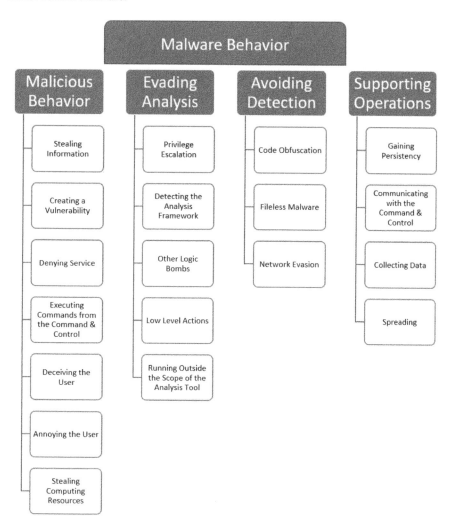

Figure 1.3. *Malware behavior (Or-Meir et al. 2019)*

1.1.2.1. *The special case of viruses*

The use of the term "computer virus", first used by Fred Cohen and Len Adleman in 1983, became widespread in the early 1990s. Computer viruses belong to the vast field of computer risks. If we go through the tree of computer risks, we see that computer viruses and worms belong to the category of self-replicating programs, which itself derives from the category of computer infections, logical attacks and finally malicious intent.

The general definition of a virus was given by Fred Cohen in 1986 as part of his doctoral thesis *Computer Viruses*.

DEFINITION (VIRUS).–

A virus is a sequence of symbols which, when interpreted in a given suitable environment, modifies other sequences of symbols in that environment so as to include a copy of itself, which may have evolved.

Thus, there are four criteria to distinguish a computer virus from another program:

1) a virus is specific to a given environment;

2) a virus has the sole function of multiplying;

3) a virus alone is inert;

4) the virus can only multiply in the host environment for which it is defined.

A computer virus therefore uses the host environment for which it is defined, such as a computer, a network, an operating system or a specific application, to multiply and infect other entities in the same environment. Fred Cohen uses the formalism of the Turing machine to define the concept of a viral set.

DEFINITION (VIRAL SET).– (Filliol 2004; Filiol et al. 2006)

For all Turing machines M and all non-empty sets or Turing programs V, the pair (M,V) is a viral set if and only if, for each virus v ∈ V and for all histories of the machine M:

– For all time instants t ∈ N and cells j of M:

1) the tape head is in front of cell j at time instant t;

2) M is in its initial state at time instant t;

3) the tape cells starting at index j hold the virus v; then, there exists a virus v' ∈ V at time instant t' > t and at index j' such that:

i) index j' is far enough from the v position (start location j);

ii) the tape cells starting at index j' hold the virus v';

iii) at some time instant t'' such that t < t'' < t' , v' is written by M.

To summarize, we can write that V is a viral set with respect to M if and only if $[(M,V) \in \mathcal{V}]$ and that v is a virus with respect to M if and only if $[v \in V]$ such that $[(M,V) \in \mathcal{V}]$.

The Turing machine M is in its initial state HM (0) at time t. It then points to the cell j of the tape containing the start of a Turing viral program v.

Cohen's (1987) formal definition indicates that at time t' > t, a new Turing viral program v' (whose start is placed in the cell j' >> j of the tape) was written by M at the time interval t' - t.

Formal definition of a viral set $\mathcal{V}$:

$\forall M \ \forall V \ (M, V \ ) \in \mathcal{V} \Leftrightarrow [V \in TS]$ and $[M \in \mathcal{M}]$ and
$\qquad [\forall v \in V \ [\forall HM \ [\forall t \ \forall j$
$\qquad\qquad\qquad [\ 1. PM(t) = j$ and
$\qquad\qquad\qquad 2. E_M(t) = E_M(0)$ and
$\qquad\qquad\qquad 3. (\square_M(t, \ j), \ . \ . \ . \ , \ \square_M(t, \ j + |v| - 1)) = v]$

$\qquad \Rightarrow \qquad [\exists v' \in V \ [\exists t' > \ t [\exists j'$
$\qquad\qquad\qquad [\ 1. \ [[(j'+ |v'|) \le j]$ or $[(j + |v|) \le j']]$
$\qquad\qquad\qquad 2. (\square_M(t',j') \ , \ . \ . \ . \ , \ \square_M(t', j' + |v'| - 1)) = v'$ and
$\qquad\qquad\qquad 3. [\exists t'' \ such \ that \ [t \ < \ t'' < \ t'\ ]$ and
$\qquad\qquad\qquad\qquad [PM(t'') \in j' \ , \ . \ . \ . \ , \ j' + |v'| - 1]$
]]]]]]]]

where:

– the "state(time)" function $E_M: N \rightarrow S_M$, which maps a move to the state of the Turing machine after that move;

– the "tape-contents (time, cell number)" $\square_M : N \times N \rightarrow I_M$, which maps a move and a cell number (cell index) on the infinite tape to the tape symbol on that cell after that move;

– the "cell(time)" $P_M: N \rightarrow N$, which maps a move to the number of the cell in front of the tape head after that move.

The Turing machine $M = (S_M, I_M, 0_M, N_M, D_M)$ is defined by giving:

- a set of n + 1 states S_M = {s0, s1, . . . , sn} with n ∈ N;

- a set of m + 1 symbols I_M = {i0, i1, . . . , im} with m ∈ N;

- a set d = {−1, 0, +1} of the possible tape motions;

- an output function O_M: S_M × I_M → I_M ;

- a state transition function N_M: S_M × I_M → S_M;

- a motion function D_M: S_M × I_M → d;

- the Turing machine history: $H_M(t)$ = (E_M, $\square_M$, P_M)(t) = ($E_M(t)$, $\square_M$ (t, i), $P_M(t)$) i ∈ N.

Cohen's definition of a virus describes a program whose only function is to copy itself. The possible malicious load carried by the virus is not included in this definition. It is only the capacity for self-replication that constitutes the virus in this formal definition. The replication capacity of computer viruses is widely used by attackers to carry the malicious loads that will impact targeted systems.

The following theorem proves that absolute viral detection is a "mathematical impossibility".

THEOREM 1.1. (UNDECIDABILITY OF VIRAL DETECTION – FRED COHEN).– (Cohen 1987; Adleman 1988; Bonfante et al. 2006)

[$\not\exists D \in \mathcal{M}$ $\exists s_i$ ∈ S_D such that $\forall M \in \mathcal{M}$, $\forall V \subset I^*$

1) D halts at a time instant t;

2) [$S_D(t)$ = s_i] ⟺ [(M, V) ∈ $\mathcal{V}$]].

PROOF.–

The proof of this theorem is mainly based on the reduction from the halting problem, which is itself undecidable. The broad lines of the proof are as follows:

1) we take an arbitrary machine M' and a tape sequence v';

2) we generate a machine M and a sequence v performing the following actions:

i) copy v' from v;

i) simulate the execution of M' on v';

iii) if v' halts on machine M', replicate v.

Thus, v replicates itself if and only if sequence v' would halt on machine M'.

Since the halting problem is undecidable and since any program which is capable of self-replication is a virus, whether $[(M, \{v\}) \in V]$ is undecidable too.

This theorem remains one of the most important results of Cohen's (1987) thesis. It limits the ambition of universality in all the antivirus solutions put on the market by the editors and imposes on them a form of humility in the face of the complexity of detection.

1.1.2.2. *The general case of malware*

1.1.2.2.1. How to formally define the set of all malware?

Giving a universal formal definition remains an open and complex problem. This definition must be applicable to all categories of malware (botnets, cryptolockers, cryptominers, rootkits, Trojans, viruses, worms) while respecting the specific actions of each malware.

In 2010, Kramer and Bradfield (2009, 2010; Bradfield and Stirling 2007) constructed the formal definitions of "Damaging Software", "Repairing software", "Malware Logic" and "Benware" from Knaster-Tarski's fixed-point theorem.

DEFINITION (DAMAGING SOFTWARE).– (Kramer and Bradfield 2009)

A software system S damages a correct software system S' by definition if and only if S (directly or indirectly) causes incorrectness in S'.

Formally:

S damages S': iff we have correct (S') and not correct (S(S')) (directly);

S damages ^{0} S': iff S damages S' (indirectly);

S damages $^{n+1}$ S': iff there is S'' s.t. not S'' damages ^{0} S' and S(S'') damages ^{n} S';

S damages ^{0} S': iff $U_{n \in N}$ S damages ^{n} S'.

DEFINITION (REPAIRING SOFTWARE).– (Kramer and Bradfield 2009)

A software system S repairs an incorrect software system S' by definition if and only if it (directly or indirectly) causes correctness in S'. Formally:

S repairs S': iff we have not correct (S') and correct (S(S')) (directly);

S repairs ^{0} S': iff S repairs S' (indirectly);

S repairs $^{n+1}$ S': iff there is S'' s.t. not S'' repairs ^{0} S' and S(S'') repairs ^{n} S';

S repairs ^{0} S': iff $U_{n \in N}$ S repairs ^{n} S'.

DEFINITION (MALWARE).– (Kramer and Bradfield 2009)

A software system S is malware by definition if and only if S damages non-damaging software systems or software systems that damage malware.

The definition of malware is iterative: everything is malware except for the following systems:

0) non-damaging systems (CP);

1) systems that damage only systems that damage CP (ATF1);

2) systems that damage only systems that damage ATF1 (ATF2);

3) systems that damage only systems that damage ATF2 (ATF3);

4) etc.

DEFINITION (BENWARE).– (Kramer and Bradfield 2009)

A software system S is benware by definition if and only if S is non-damaging or damages only software systems that damage benware.

The definition of benware is iterative: nothing is benware except for the following systems:

0) non-damaging systems (CP);

1) systems that damage only systems that damage CP (ATF1);

2) systems that damage only systems that damage ATF1 (ATF2);

3) systems that damage only systems that damage ATF2 (ATF3);

4) etc.

1.1.2.2.2. The malware versus benware arms race

The good and bad distinction is induced by the existence of a population that is (perceived as) non-damaging (Bradfield and Stirling 2007; Kramer and Bradfield 2009).

The malware–benware duality translates into the following diagram:

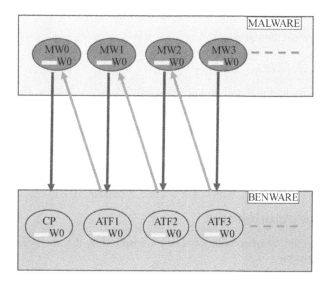

Figure 1.4. *Malware versus benware arms race. For a color version of this figure, see www.iste.co.uk/briffaut/complexities1.zip*

1.1.3 *Cyber-risk*

Cyber-risk is a relatively new risk compared to other types of risks, such as fire risk. But it is an exponential risk. Since an attack can affect from one to several million interconnected computers in many countries around the world, the reinsurance industry believes that cyber-risk is as important as the risk of natural disasters. It was estimated at 600 billion dollars (€534.4 bn) in 2018.

1.1.3.1. *Cyber-risk definition*

For any legal or natural person, referred to as "the entity", cyber-risk refers to all attacks on:

– electronic and/or computer systems (for the production, operation, and management of information and telecommunications) under the control of the entity or its service providers;

– computerized data (personal, confidential or operating data) belonging to or under the control of the entity, whether transferred or stored with it or its service providers.

It can be consecutive to:

– a malicious or terrorist act;

– human error, breakdown or technical problems;

– a natural or accidental event.

As a result, we have:

– bodily injury and material and/or immaterial damage (costs or financial losses) suffered by the entity and/or its employees;

– mobilization of internal or external resources – personal injury, material and/or immaterial damage, costs or financial losses caused by the entity to third parties (including supply chain/subcontractors);

– an attack on the brand and/or reputation of the entity.

1.1.3.2. *Formal global definition*

A high-level notion of conventional risk is inspired by the management literature (Böhme et al. 2018):

Risk = probability of a loss event × magnitude of loss

This definition of risk has many deficits. It coerces a complicated loss distribution to a single Bernoulli trial, it is agnostic about the time dimension and it does not differentiate between individual and aggregated losses caused by a single loss event.

Malicious actors	Motivations	Targets	Impacts
Nation	– Economical, political and military	– Secrets, patents – Sensitive information – New/emerging technology – Critical infrastructure – Strategic suppliers	– Loss of advantage and competitiveness – Disruption of critical infrastructure – Political/diplomatic conflicts

Malicious actors	Motivations	Targets	Impacts
Criminal organization	– Immediate financial gain – Gathering information for future financial gain	– Financial settlement/payment systems/cash terminal servers – Payment card information (PCI) – Identifiable employee information – Information/customer data and credit cards	– Expensive regulatory investigations and sanctions – Lawsuits brought by customers and shareholders – Loss of consumer confidence
Terrorist organization	– Political and ideological or religious in order to create general disorganization, as well as fear and panic	– All information systems: media, companies, public authorities, government sites, SCADA systems and private individuals	Threatens peace and national security through actions with consequence of: – property damage – personal injury
Hacktivist	– Political influence and/or social/societal change – Pressure on companies to change their practices	– Industrial or trade secrets – Sensitive commercial information/pricing policy/supplier references – Information on key management, staff, customers and partners	– Disruption of business operations, loss of markets – Brand and reputation – Loss of consumer confidence
Insiders	– Personal benefit, financial gain – Professional revenge – Patriotism	– Sales, offers, strategic markets – Industrial and commercial secrets, including partners and to a lesser extent patents, R&D – Operations, commercial strategies – Personal information	– Disclosure of trade secrets, patents – Disruption of operations – Brand and reputation – Impact on safety

Table 1.2. *Cyber-risks of malicious/criminal origin*

Accidental incidents	Nature of the event	Consequences	Financial impacts
Human error	– Programming error – Implementation error – Error of use – Maintenance error	– Shutdown of information systems – Liability in case of errors in customer operation, delivery of goods or services – Loss or alteration of customer data, confidential data or operating data – Regulatory procedure	– Additional operating costs – Defense costs/damages – Data reconstitution fees – Operating losses – Cost of replacement equipment – Fees and administrative penalty
Breakdown and technical difficulties	– Maintenance fault – Problem of putting a software into production – Problem of interoperability of systems with suppliers, third parties, customers – Industrial electrical origin: contact switching, operation of thyristors, etc. – Electronic: distribution network and relay problems		
Natural event	– Fire – Flooding – Water damage – Storm – Lightning and related electrical surges		

Table 1.3. *Cyber-risks of accidental origin*

1.1.3.3. *Risk of damage to the data held, collected, hosted and used*

Data security must meet the following three main objectives:

1) *Availability*: information must be accessible to all who need it (and who are authorized to have it).

2) *Confidentiality*: the information must remain accessible only to authorized persons.

3) *Integrity*: information must not be corrupted or made incomplete.

The following two further objectives broaden the previous ones to define the security of the information system:

1) *Non-repudiation and imputation*: no user should be able to challenge the actions of another user, and no third party should be able to attribute the actions of another user to themselves.

2) *Authentication:* user identification is fundamental to manage access to the system in relevant workspaces and maintain confidence in the exchange relationships.

A company's information system uses two types of data: company-owned data and third-party data. These two sets are made up of heterogeneous data: personal data, confidential data, operating data, regulated data, customer data, personal data, industrial secrets, trade secrets, R&D, Big Data, competitive data, supplier referencing and pricing policy, strategic and financial data, brands, archive data, logistic data, consumer data, employee data, HR data, etc.

1.1.3.4. *Cyber-risks and financial impacts*

A company's information capital consists of:

– primary assets: processes and activity, information;

– support assets: hardware, software, networks, personnel, sites, organizational support (regulatory authorities, parent company, departments, agencies, etc.).

When assessing a company's cyber-risk, a distinction must be made between the losses the company incurs and those it causes to third parties.

The cyber-risk matrix relates the probabilities of cyber-threats to the financial consequences of these attacks for the targeted entity (company or administration).

Losses incurred	Potential financial impacts
Loss or damage to data	Loss or deterioration of data or software, resulting in costs to restore, update, reconstitute or replace these assets

Business interruption or network unavailability	Operating loss in the event of an interruption, degradation of service or network slowdown resulting in loss of revenue, increased operating costs and/or mitigation and investigation costs
Damage to reputation	Violation of data protection that results in loss of intellectual property, loss of revenue and loss of market share
Regulator's investigation of privacy breaches	Investigation, regulatory procedure (GDPR); defense costs, fines resulting from an investigation
Notification fees	Legal, postage and communication costs in countries where there is a legal or regulatory obligation to inform individuals of a breach of security or confidentiality, including associated reputational costs

Table 1.4. *Cyber-risk and financial impact*

Likelihood	Consequences				
	Insignificant (minor problem easily handled by normal day to day processes)	**Minor** (some disruption possible, e.g. damage equal to $500,000)	**Moderate** (significant time/resource required, e.g. damage equal to $1 million)	**Major** (operations severely damaged, e.g. damage equal to $10 million)	**Catastrophic** (business survival is at risk, e.g. damage equal to $25 million)
Almost certain (e.g. >90% chance)	High	High	Extreme	Extreme	Extreme
Likely (e.g. between 50% and 90% chance)	Moderate	High	High	Extreme	Extreme
Moderate (e.g. between 10% and 50% chance)	Low	Moderate	High	Extreme	Extreme

Unlikely (e.g. between 3% and 10% chance)	Low	Low	Moderate	High	Extreme
Rare (e.g. < 3% chance)	Low	Low	Moderate	High	High

Table 1.5. *Cybersecurity risk matrix. The colors correspond to the priority level of the problem to be taken into account. Red: extreme security priority; yellow: high security priority; dark beige: moderate security priority; light beige: low security priority. For a color version of this figure, see www.iste.co.uk/briffaut/complexities1.zip*

1.1.4. *Cognitive attacks and immersive fictitious data architectures*

Cognitive attacks target a human user or operator with access to a civilian or military computer system. Their objective is to make him/her execute actions against his/her interest by exploiting his/her cognitive biases and weaknesses. They are generally based on fictitious data architectures that are more or less immersive and are designed to install trust in the target and then deceive him/her in his/her future actions on the system.

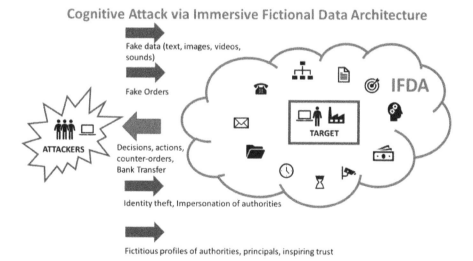

Figure 1.5. *Mechanism of an IFDA attack. For a color version of this figure, see www.iste.co.uk/briffaut/complexities1.zip*

Artificial intelligence (AI) facilitates the construction of immersive fictitious data architectures (IFDAs). IFDAs can be used to deceive a human target, an industrial system operator (Viveros 2016), a company's accounting manager, a political authority or a military leader. The immersive nature of an IFDA builds trust with the target and induces him to perform actions in favor of the attacker or to facilitate the transfer of information to the attacker. The social engineering phase of a targeted cyberattack increasingly relies on sophisticated and persistent IFDAs over time. These need to be kept consistent, credible and immersive throughout the operation.

1.1.4.1. *Cognitive attack modeling via an immersive fictional data architecture*

1.1.4.1.1. The attacker and his/her sequence of actions

Attacker A is motivated by an objective to be achieved, by a strategy S to be implemented and by a winning function to be maximized.

The attacker executes a sequence of actions:

$$[(AR_1, AD_1), (AR_2, AD_2), \ldots\ldots, (AR_n, AD_n) \rightarrow \text{objective achieved or not}]$$

where AR_i is an action performed in physical space and where AD_i is an action performed in the digital space (sending messages, datasets, malicious or inert files, links, videos, audio or text, running programs on a compromised machine).

The actions of the attacker (AR_k, AD_k) executed at level k are decided (calculated) according to the history of the past actions of the attacker, their consequences, the objective to be reached and the actions of the target T executed in the physical and digital spaces: (TR_i, TD_i), $i = 1,2, k-1$

This is how we have:

$$(AR_2, AD_2) = S < (AR_1, AD_1); (TR_1, TD_1) >$$

$$(AR_3, AD_3) = S < [(AR_1, AD_1), (TR_1, TD_1), (AR_2, AD_2)]; (TR_2, TD_2) >$$

.....

$$(AR_k, AD_k) = S < [(AR_1, AD_1), (TR_1, TD_1), \ldots\ldots, (AR_{k-1}, AD_{k-1})]; (TR_{k-1}, TD_{k-1}) >$$

The attacker must maintain the target's confidence and belief in the veracity and integrity of the sequence:

TRUST-TARGET < [(AR$_1$, AD$_1$), (TR$_1$, TD$_1$), ... , (AR$_{k-1}$, AD$_{k-1}$), (TR$_{k-1}$, TD$_{k-1}$), (AR$_k$, AD$_k$)] > = 1 (if 0 stop)

The attacker must preserve the non-contradiction (consistency) of the set sequence:

CONSISTENCY < [(AR$_1$, AD$_1$), (TR$_1$,TD$_1$), ... , (AR$_{k-1}$, AD$_{k-1}$), (TR$_{k-1}$, TD$_{k-1}$), (AR$_k$, AD$_k$)] > = 1 (if 0 stop)

The attacker must maximize the immersive nature of the sequence.

1.1.4.1.2. The target and its actions

The target produces a series of actions in physical and digital space [(TR$_1$, TD$_1$), (TR$_2$, TD$_2$), ... , (TR$_n$, TD$_n$), ...] based on its usual activity and in response to the attacker's sequence of actions.

1.1.4.1.3. The end of the sequence

The attack sequence stops when the attacker's objective has been reached (via its gain function) or when the target's confidence level cancels: TRUST-TARGET = 0 with or without CONSISTENCY = 0.

Figure 1.6. *Fake GAN video of Barack Obama (BuzzFeedVideo 2018)*

IFDAs contain text, sounds, images and videos. They can be based on fictitious profiles built from photos of fictitious people created by a generative adversarial network (GAN) (Goodfellow et al. 2014). Deep learning technologies make it possible to mimic an individual's fingerprints or voiceprints. They are also used to create videos of military or political authorities delivering fictitious speeches that are perfectly realistic and undetectable (BuzzFeedVideo 2018). AI can thus generate credible false information (deep fakes).

Figure 1.7. *Fake person – GAN (source: "This person does not exist" (2019); https://www.thispersondoesnotexist.com/)*

Figure 1.8. *Fake cat – GAN (source: "This cat does not exist" (2019); https://thiscatdoesnotexist.com/)*

AI can be used to build a cyberattack. In 2018, an IBM research team (Kirat et al. 2018) demonstrated that it is possible to encapsulate malware in a software layer that includes machine learning components.

1.1.4.2. *AI will be an effective shield against new threats*

Some of the threats described above can be countered by security solutions that incorporate machine learning components. This is the case in the detection of targeted cyberattacks whose objective is the exfiltration of sensitive data. User behavior analytics (UBA) uses automatic learning techniques to learn the normal operation of a network and then to detect anomalies by cross-referencing a very large amount of data produced by the information system. Natural language processing (NLP) is also very useful for detecting fraud and certain fictitious data structures.

AI will make it possible to test the consistency and veracity of a corpus of data while seeking to determine its origin. Coupled with blockchain technologies, it contributes to certifying a set of data by giving it traceability.

The time factor remains central during the detection of intrusion attempts in a system. AI provides efficient responses for automatic detection of high frequency incidents. It also intervenes downstream of the attack in the incident response phase by guaranteeing automatic reconfiguration of the system in degraded mode. Finally, formal computation methods are used in code proofing processes as early as the development phase of a program. They guarantee security "by design".

In sections 1.2 and 1.3, we show how machine learning techniques improve the detection, classification and handling of cyber-threats. By analyzing all the data coming back from the information system, UBA tools use all the interactions of the "human–machine–software" system to measure, in real time, the level of potential threats. In this sense, they are tools that moderate the system's cybersecurity complexity.

1.2. Cybersecurity in real life: Advanced persistent threats, computer networks, defense teams and complex log data

1.2.1. *What is an APT?*

Until a few years ago, cyberdefense mainly consisted of identifying the main threats to organizations and finding the appropriate antivirus. As the number and variety of cyberattacks have been constantly growing (Jang-Jaccard and Nepal 2014) over the last few years, things have changed a lot. The landscape of cyber-attacks has become extremely large and complex. Today, the new threats in cyber-space for major companies are called advanced persistent threats (APTs).

What exactly is an APT? The term APT was introduced in 2006 by the US Air Force defense service in order to describe "specific opponents, exploits and targets designed to steal strategic data". In other terms, an APT is a *custom attack designed to hit a specific target with a precise objective* (stealing sensitive data, blocking the activity of a cyberphysical system[1], asking for a ransom, etc.). This means two things:

– the group of individuals behind the APT is organized and has the financial resources required to schedule the attack[2];

– the attack is customized and designed to fade into the network's background (over several years if needed) until the impact phase. It is therefore extremely difficult to *detect* and *block* it.

We show later in this chapter that *detecting an APT can be seen as dealing with the data generated by the informatic network and reducing its complexity using human and algorithmic resources.*

Every APT attack can be broken down into (at least) three stages:

1) network infiltration (through phishing, for instance);

2) expansion (gaining access to a privileged user account, like a system administrator);

3) impact (extracting data, perturbating[3] or even blocking the target network).

There are a lot of different possibilities for the hackers at every stage. For instance, the most common way of performing network infiltration is social engineering and phishing, but it can also be done using the vulnerability of a public-facing application[4].

---

1 Cyberphysical systems are referred to as situational crime prevention (SCP) systems in the rest of this chapter.

2 Review of the main presumed APT groups: https://www.fireeye.fr/current-threats/apt-groups.html.

3 In 2015, the APT28 group managed to disrupt the French TV channel *TV5 Monde* and display an Islamic propaganda message.

4 In cybersecurity terms, unknown computer-software vulnerabilities exploited by hackers to penetrate systems are called 0-day vulnerabilities.

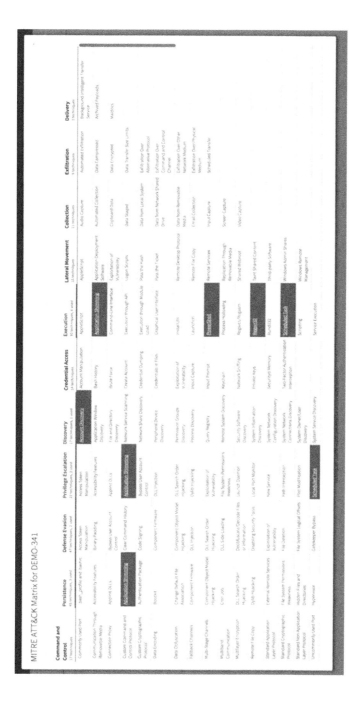

Figure 1.9. *Representation of an APT in the MITRE ATT&CK framework. Red columns represent the tactics (what has been done) and red cells represent the technics (how it has been done). For a color version of this figure, see www.iste.co.uk/briffaut/complexities1.zip*

The most significant contribution breaking down all the techniques used by APTs is the MITRE ATT&ACK matrix[5]. It is a comprehensive framework listing all the tactics and technics used by APT groups at every stage of the attack:

– A tactic is a step in the attack (initial access, privilege escalation, exfiltration, etc.). It corresponds to the "what".

– A technic is way of performing this step and moving on to the next one. It corresponds to the "how".

The overall defense process of detecting and blocking an advanced threat is composed of subprocesses, linked to the detection of every step of the APT. Once all the red cells of the MITRE ATT&CK matrix have been identified, the forensic team perform its analysis and is able to tell if the sequence of steps matches the signature of a known APT.

Generally speaking, we can see the process of detecting advanced threats as a *two-step complexity reduction process* using the data generated by the network assets as input:

– point-in-time complexity reduction: detecting anomalies in the network at specific moments;

– temporal complexity reduction: correlating all the anomalies to reconstruct the attack.

1.2.2. *What is the network that companies need to protect? Who protects it? Why are "Situational Crime Prevention" (SCP) systems complex systems?*

A computer network is a digital telecommunications network that allows nodes to share resources. In computer networks, computing devices exchange data with each other using connections (data links) between nodes. The nodes in the network are identified by IP addresses. Computer networks also interact with the outside world (the World Wide Web) through an enormous number of applications and services. The interaction between the network and the outside world is managed through a set of assets, ports[6] and protocols.

---

5 https://attack.mitre.org/.

6 Traditionally, port 25 is used for e-mails (SMTP protocol), port 80 for HTTP, port 443 for secured HTTP, etc. The list of the most commonly used ports can be found here: https://en. wikipedia.org/wiki/Port_(computer_networking).

1.2.2.1. *Main risk factors*

The protocol and devices facing the highest cybersecurity risks are as follows:

– The DNS[7]: it translates every internet request (like accessing www.google. com) to a physical IP address in the network (for instance, 8.8.8.8 for Google). As there is a lot of data transiting through the DNS in both directions (from the inside to the outside, and vice versa), it is a privileged target for hackers.

– The *firewall* controls network security and access rules. It is configured to reject access requests from unrecognized sources while allowing actions from recognized ones. It plays a vital role in network security as it manages all the messages exchanged in the network and toward the outside world. Firewalls are able to blacklist well-known malicious sources and to apply static security rules. The challenge for APT is to design an intrusion method with sufficient complexity to bypass the static rules. For instance, an alert consisting of blocking messages sized greater than 100 bytes over port 53 can be bypassed by sending three messages of 40 bytes on this port, through three different sources. This can be done through *botnet command and control*[8], where a command server takes control of bots and tells them where to send all the individual messages.

– The *proxy* is a server application that acts as an intermediary for requests from clients seeking resources from servers that provide these resources. The proxy evaluates the request and performs the required network transactions. It is an additional layer of defense, but it is exposed to attacks from the Internet[9].

– The *active directory* is the means by which users, customers, partners, the IoT and other edge devices authenticate a system and receive their rights for traversing that system. It is an important target[10] as it might be used by cybercriminals to:

- discover the network and the users (infiltration phase);

- move laterally in order to carry out multistage attacks (expansion phase);

- gain access and abuse strategic data (impact phase).

Throughout time, users in computer networks interact with each other and generate data through the different assets mentioned above. These log data can be

---

7 For instance, DNS tunneling is one of the most popular cyberattacks: https://www. paloaltonetworks.com/cyberpedia/what-is-dns-tunneling.

8 See: https://attack.mitre.org/techniques/T1041/.

9 For instance, see: https://attack.mitre.org/techniques/T1090/.

10 List of the techniques mitigated by AD configuration: https://attack.mitre.org/mitigations/ M1015/.

stored in the security information event management (SIEM) system. They can be used by dedicated teams to perform complexity reductions in order to detect advanced threats.

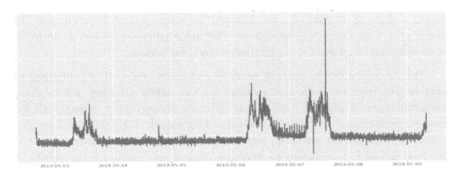

Figure 1.10. *Number of nodes in a firewall network over one week*

NOTE ON FIGURE 1.10.– *Temporal evolution of the number of IP addresses sending messages through the firewall between May 2 and May 9. We can see some patterns (low number of nodes at night and during the weekend, high number of nodes in the morning and the afternoon) and some irregularities (sudden rises or falls during the day).*

1.2.2.2. Teams involved in cyberdefense

The security operation center (SOC) is in charge of protecting the network against the rising number of cyber-threats. It is composed of:

– *Security operation center analysts*: the SOC monitors all the information generated by the assets described in the previous section and by data centers, servers and all the endpoints of the network. The analyst's main task is to detect signs of a possible cyberattack or intrusion by monitoring these systems actively. The biggest challenge is to deal with the huge volume of data in real time and provide fast and precise cyber-threat analysis and reports.

– *Computer emergency response teams* (CERTs) are the people handling all the security incidents. Their role is to coordinate and execute response strategies in order to contain the attack and remediate to it. They have to deal with a very high number of false positives and take actions proportionate to the severity of the threat.

– *Digital forensic experts*: after the attack has happened, these experts are in charge of:

- investigation in order to collect the evidence (files, phishing e-mails, etc.);

- identification of the attacker;

- selection of appropriate countermeasures.

– *Red teams* and penetration testers (pentesters) help enterprises to improve themselves by providing opposition. They are focused on penetration testing of different systems and their security programs. They detect, prevent and eliminate vulnerabilities. The red teams imitate real attack tactics and technics.

– The *blue team's* role is also to assess the network security and to identify potential vulnerabilities. It does not simulate any attacks but rather finds ways to defend, change and re-group defense mechanisms to make incident response stronger. The classical steps a blue team incorporates are security audits, digital footprint analysis, reverse engineering and scenario testing.

1.2.2.3. *SOC analysts have to deal with complex data*

The logs generated by the network's assets can take a lot of different forms. We can regroup the logs into five categories:

– *Endpoint logs*: an endpoint is a computing device within a network, such as a laptop or smartphone. Endpoints generate multiple logs from different levels of their software stack – hardware, operating system, middleware and database application. They can be used to understand the status activity and health of the endpoint device.

– *Firewall and router logs*: network devices like firewalls, routers, switches and load balancers are the backbone of the network infrastructure. Router logs provide information about traffic flows (websites visited by internal users, sources of external traffic, protocols used, etc.).

– *Application event logs*: applications running on servers or end user devices generate and log events with information like startup, shutdown, heartbeat and runtime error events from running applications. Example of these logs can also be email, Web or database servers.

– *IoT logs*: the IoT is a new and growing source of log data. IoT deployments save log data to a central cloud service.

– *Proxy logs*: proxy server logs contain requests made by users and applications on a local network, as well as application service requests made over the Internet such as application updates.

```
ts;src;dst;port;sent_bytes
2019-06-25 07:10:31.529000+00:00;192.168.120.1;172.22.8.42;22;292.0
2019-06-26 06:22:22.667000+00:00;192.168.120.1;10.0.3.10;22;292.0
2019-06-27 01:02:33.439000+00:00;192.168.120.12;192.168.140.1;53;73.0
2019-06-25 07:40:19.335000+00:00;192.168.120.19;172.217.171.195;443;1788.0
2019-06-26 09:38:41.320000+00:00;192.168.150.1;192.168.140.1;53;61.0
2019-06-26 06:22:23.797000+00:00;192.168.150.106;8.8.8.8;53;71.0
2019-06-27 01:02:33.518000+00:00;192.168.120.1;10.177.40.160;443;284.0
2019-06-25 07:40:19.975000+00:00;192.168.120.6;192.168.140.1;53;146.0
2019-06-26 09:38:41.320000+00:00;192.168.150.1;172.217.19.142;443;1445.0
2019-06-25 07:10:31.730000+00:00;192.168.120.7;18.202.184.175;443;164.0
```

Figure 1.11. *Examples of firewall log messages*

NOTE ON FIGURE 1.11.– *First 10 rows of a firewall log file. Each line contains one message described by the timestamp (ts), the source (src) and destination (dst) IP addresses, the port of communication (port) used, and the number of bytes sent (sent_bytes). Firewall logs can reach up to 40 million rows per day!*

At the end of the day, the cybersecurity logs are stored in the SIEM. They contain collections of *multivariate event data* generated by many different types of assets. Each event stored in the logs is characterized by its time of occurrence, its type and additional properties called metadata. For instance, the first row of the firewall log above is a message containing 292 bytes of information. It was sent from IP 192.168.120.1 to IP 172.22.8.42 on port 22 on June 25, 2019 at 07:10:31 AM. The metadata are not always available in the logs. In this case, examples of metadata could be the owner of the device registered with IP address 192.168.120.1, the nature of the work that this person was doing when the message was sent, etc.

The cybersecurity logs are characterized by:

– *Size*: for computer networks consisting of hundreds or thousands of nodes, the amount of communications easily reaches millions of events per second.

– *Nature*: most of the data are numeric (for instance, firewall logs contain numeric NetFlow[11] data), but they might be textual (e-mails), hexadecimal, compiled code, etc.

– *Variety*: the number of attributes and amount of metadata can easily reach hundreds of attributes and more.

As a result, systems are extremely heterogenic and have high numbers of dimensions, which makes human exploration very difficult.

---

11 NetFlow is a feature introduced by Cisco in 1996 to collect IP network traffic.

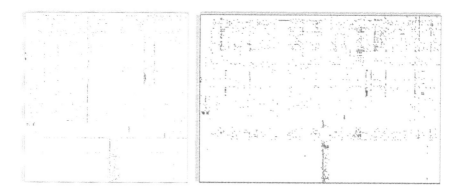

Figure 1.12. *Firewall data complexity pattern*
(at 2 AM on the left and 9 AM on the right)

NOTE ON FIGURE 1.12.– *Complexity analysis of a firewall network. Both pictures represent 1 min of firewall communications. Source IP addresses are displayed as rows and destination IP addresses as columns. Every blue dot represents one instance of communication between a source and a destination. It is obviously very difficult to establish clear patterns in the communications. Both matrixes are very sparse (the majority of the dots are white, which means that most of the IPs do not communicate with each other). There are more communications at 9 AM, and some IP addresses communicate more than others: the vertical line at the bottom means that several sources send messages to one destination. In the right picture, there are some horizontal lines (one source sending messages to a lot of different destination IPs).*

1.2.2.4. *SCPs are complex and adaptative*

Cybersecurity logs capture all the events and relations of the cyberphysical systems. As the data in the logs are highly complex, SCPs can be described as *complex and adaptative*. A complex system is a system in which the parts of the system and their interaction together represent a specific behavior which cannot be explained as the analysis of all its constituents' parts: "the whole is greater than the sum of its parts". This behavior can be described by the following three properties:

– the overall behavior cannot be predicted simply by analyzing the parts and inputs to the system (as SCPs integrate highly unpredictable human agents, it is impossible to predict the future state of the system);

– the behavior of the system changes with time: the same input and external conditions do not always guarantee the same output behavior (for instance, a laptop might be redeployed to another individual within the organization);

– the agents of the system change their behavior based on the outcome of previous experience. In an SCP, it is not possible to relate all causes and consequences.

The challenge when analyzing an SCP is to detect when it is being exploited; this means finding a message or a set of messages that are corrupted or malicious from the whole set of information exchanged in the network. Every message contains data and metadata. The main difficulty lies in the fact that it is impossible to tell if a message is malicious only by looking at the data. One also need to look at:

– Who sent the message: during the infiltration phase, the malicious source is outside the network (most of the time, this source is the server or the client sending the phishing message). During further steps, like lateral movement or exfiltration, the source is not active anymore and the malicious messages are sent by a corrupted machine located inside the network.

– The exact content of the message (deep packet inspection[12]).

– Who received the message.

– When the message was sent: advanced cyberattacks might last several months or several years; so we would need to compare every message of the network with all other messages exchanged during the last years in order to be able to tell if it is normal or not.

– How the message modified the network or the behavior of specific entities in the network.

As a result, it is almost impossible for a team of SOC analysts to discover all the patterns and anomalies in the logs. The data are a combination of *multivariate network data* and *temporal data* and not humanly interpretable. In order to reduce complexity and allow human analysis, specific events considered as suspicious must be isolated and processed by SOC analysts.

1.2.3. *What kind of anomalies need to be raised in order to detect a multi-stage APT attack?*

In order to process high-dimensional network-generated data and prevent cyber-threats, security teams need to perform log analysis and raise an anomaly as soon as the observed behavior seems to be suspicious. This means that all the events need to be analyzed and one "probability of anomaly" has to be attributed to each event. The overall log analysis task can be seen as the discovery and understanding

---

12 Deep packet inspection is a type of data processing that inspects in detail the data being sent over a computer network.

of patterns and anomalies. There are lots of different ways of defining an event as anomalous. For instance, two IP addresses exchanging a lot of data can be anomalous. But how should we define "a lot of data"? On the other hand, two IP addresses without any communications can also be anomalous. But how do we tell if these IP addresses were supposed to exchange data? Does it depend on the weekday number? The hour of the day? Should we raise an anomaly if somebody in the network downloads a movie (as the size of a movie file corresponds to the size of thousands of e-mails, a movie is "a lot of data")? If we do so, the number of false positives will be extremely high, and we face the risk of overloading the computing emergency response team (CERT), which will probably miss the "real" cyber-attack by losing time to investigate these anomalies.

The following sections provide answers to these questions by categorizing the anomalies into two different classes: *point-in-time* and *temporal correlation*. We show that a real cyber-threat has to be characterized by the combination of these two categories.

1.2.3.1. *Point-in-time anomalies*

Point-in-time anomalies are events that are anomalous with respect to the entire dataset. They are unusual irrespective of the context in which they are observed. From the statistical point of view, point-in-time anomalies correspond to items in the dataset which differ significantly from the majority of the data. Examples of point-in-time anomalies are as follows:

– at the network node level: an IP address downloading too much data in a short period of time;

– at the network edge level: two IP addresses exchanging too much data in a short period of time;

– at the network level: a sequence of very similar websites contacting the network in a short period of time – this is a technique used by hackers known as the domain generation algorithm: an algorithm generates domain names following a pre-defined pattern in order to organize rendezvous points for corrupted bots involved in an attack;

– at the user level: a member of the sales team exchanging a lot of data on a port of communication dedicated to secure database connections;

– at the temporal level: a short period of time where unusual activity is detected in the network (for instance, a lot of connections to a specific server during the middle of the night);

– at the entity level: a DNS sending a lot of data to a remote server.

Point-in-time anomalies are the first level of anomalies that need to be raised and analyzed. They correspond to a suspicious activity in the network but will not allow the analysts to know if an APT is targeting the network. For a security system to be effective, the number of false positives regarding point-in-time anomalies needs to be manageable by the team.

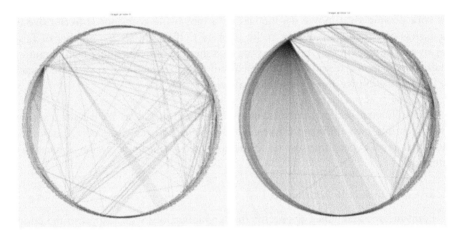

Figure 1.13. *Firewall network point-in-time anomaly*

NOTE ON FIGURE 1.13.– *Network point-in-time anomaly: the two graphs represent 3 min of firewall communication. IP addresses are represented on the circle. We draw a line if they exchange messages during the time slot. On the left, the network activity is normal. On the right, we can see that one source is sending messages to a lot of destination IP addresses. This is known as a port scan, which is a "network discovery" operation.*

1.2.3.2. Temporal correlation anomalies

Temporal correlation anomalies are collections of data points that together are considered as anomalous. This is closely related to the notion of unusual pattern. A temporal correlation anomaly involving one or many point-in-time anomalies will be characterized as an advanced threat and could potentially be an APT. Examples of temporal correlation anomalies are as follows:

– An IP address which receives a lot of messages from an external source and then sends a lot of message to another IP address in the network. It will be detected as "network intrusion followed by lateral movement", which are usually the first two steps of an APT.

– The installation of a new software version followed by point-in-time anomalies. It will be detected as a 0-day vulnerability[13], which have been exploited by hackers in order to penetrate networks and perform actions.

– A sequence of password brute force attacks over several IP addresses in the network.

The level of severity of temporal correlation anomalies depends on:

– the individual data points in the collection;

– the overall time that this anomaly lasts (APTs usually integrate a "sleep phase" between the infiltration and the execution phases).

As a result, the temporal complexity is very high because a huge combination of events needs to be analyzed in order to detect a temporal correlation anomaly.

For instance, if a network generates 100,000 events per second, this will correspond to 3153 billion events per year. Analyzing all the possible collections of 10 events would require the investigation of $\left(\begin{array}{c} 3.153 \times 10^{12} \\ 10 \end{array} \right)$ events, which is obviously impossible.

The last section of this chapter presents our novel embedding-based framework that we developed in order to facilitate anomaly detection through machine learning embedding techniques.

1.3. User and entity behavior analysis as a way of reducing complexity

1.3.1. *Presentation of the method*

Today, user and entity behavior analysis (UEBA) is the solution that companies need to use to detect anomalies in network-generated log data. Using UEBA, companies do not track security events or monitor devices; instead, they track all the users and entities in the system. They use machine learning algorithms and statistical analyses to know when there is a deviation from established patterns. When the behavior of any entity in the system changes significantly (change is defined here using advanced mathematical and statistical tools and algorithms), an alert is raised and further investigated by security analysts. AI algorithms have proven to be a very good way to learn normal entity behaviors using a variety of data sources, mainly logs (FW, AD, proxy, event logs, etc.), and accurately detect deviations from

---

13 A 0-day vulnerability is a software vulnerability that is unknown. An exploit directed at the 0 day is called a 0-day attack.

normality to raise alerts. This has led to the proliferation of machine learning and deep learning algorithms, mainly among these categories:

– recurrent neural networks (RNNs) and long-short term memory (LSTM) for temporal behavior monitoring issued from NLP (Chawla et al. 2018; Yuan and Wei 2018);

– autoencoders for anomalous user behavior and intrusion detection (Mirsky et al. 2018);

– generative adversarial networks (GANs) to create synthetic data similar to input data in order to improve algorithm performance (Sweet 2019);

– geometric deep learning and embedding methods in order to learn and predict the behavior of complex data structures like graphs and hypergraphs (Adams and Heard 2016).

These algorithms have proven to be very effective for well-defined categories of attacks, but some limitations remain:

– the deep learning algorithms need to be fine-tuned for each specific type of attack, which is extremely time- and resource-consuming;

– the (deep learning) algorithms do not tend to be tuned for a specific task and will not necessarily generalize well to new types of attacks and complex attacks like APTs;

– the algorithms are often "black boxes". As a result, mathematicians and cybersecurity analysts struggle to communicate about the results and to integrate expert feedback in the models.

In order to overcome these issues, we propose a novel embedding-based framework that facilitates UEBA by projecting sparse and unstructured log data into a new mathematical space in which numerous behavior trends and changes can be analyzed in a simpler and more visual way than when using typical deep learning algorithms. The outcome of our method is a framework in which all the entities have time-evolving geometric coordinates, reflecting the nature and the quantity of the interactions observed between these entities. This framework is used as a "dialog box" for complex attack detections: ethical hackers are able to simulate attacks and understand how they translate in this space thanks to the help of data scientists. We use "traditional" machine learning algorithms (density-based methods, clustering and neighbor analysis) that are constructed through embedding methods to track how complex attack patterns translate in this mathematical space.

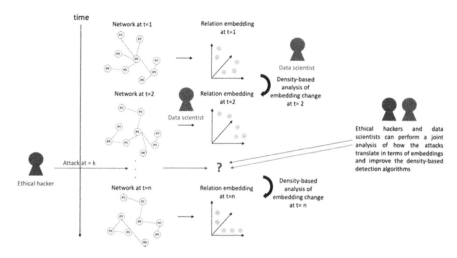

Figure 1.14. *Overview of our embedding-based detection framework. For a color version of this figure, see www.iste.co.uk/briffaut/complexities1.zip*

Embedding is one of the successful uses of deep learning in the past years. The method aims at representing discrete variables and data concepts as continuous vectors. This method works by mapping similar values close to each other in the embedding space, thus revealing the intrinsic properties of the categorical variables. It has been popularized by the great success of the Word2Vec algorithm, developed by Tomas Mikolov at Google in 2013 (Pingle 2019). This algorithm extends the skip-gram model and makes it possible to capture a large number of precise syntactic and semantic word relationships. It contributes to mitigating the "curse of dimensionality" by representing variables of interest in a denser space. The use of embedding techniques for cybersecurity is at its beginning. Research has been carried out regarding the construction of knowledge graphs (Mikolov et al. 2016).

Our method processes IP-IP communication graph data through an extension of the node2vec algorithm. We create sliding windows, and for each window the algorithm:

– encodes each type of IP behavior in an M1 multidimensional vector;

– encodes each IP–IP relation in an M2 multidimensional vector (with M2 >= M1);

– projects the IP and the relations in a two-dimensional space that makes it possible to perform density analysis in a visual way.

The proposed method is a combination of node2vec and t-SNE algorithms and will be extended in the future using autoencoder models. We refer to the work of Zhu et al. (2016) to define desired properties for the embeddings:

– temporal smoothness: IPs change their latent positions gradually over time;

– network embedding: if a couple (IP1; IP2) interacts a lot in the network, the distance between Embedding_IP1 and Embedding_IP2 will be small in the embedded space (and vice versa);

– latent homophily: IPs that are close to each other in latent space interact with one another more frequently than two faraway members.

We extend the latent homophily property to add a new constraint:

– latent relationship homophily: if the *nature* of the relation (IP1; IP2) is close to (IP3; IP4), their embeddings should be close. This should be temporally coherent.

The novelty of our method consists of the fact that our algorithm takes into account variables describing the *nature* of the relation between two IPs: port of communications, volume of data exchanged and duration of the connection. All these pieces of information are processed to compute the embedding. To our knowledge, this type of industry-wide research has never been done before.

1.3.2. *Data used and details of the method*

1.3.2.1. *Data used and simulated attacks*

Our model was tested using 17 days of firewall IP communication data (19 GB) belonging to a medium-sized company (2 days on April 15 and 16, 2019, + 15 days from May 2, 2019, to May 16, 2019). The dataset contained 80 private IPs and 9000 public IPs in the network. We used raw data with the following fields (FortiGate firewall):

– timestamp;

– source IP;

– destination IP;

– port of communication;

– number of bytes and packets sent;

– duration.

Different types of attacks were run by ethical hackers:

– ping scan (source: 192.168.120.7, victim: 192.168.140.0/24, UTC time: April 15th 3:45 pm);

– TCP port scan (source: 192.168.120.7, victim: 192.168.140.109, UTC time: April 15th 3:25 pm);

– port scans with speed lowering (source: 192.168.120.7, victim: 192.168.140.109, UTC time: from April 15th 1:56 pm to April 16th 8:07 am);

– botnet C&C and data exfiltration through the DNS (source: 192.168.120.7, DNS: 192.168.140.1, UTC time: from May 7th 12:53 pm to May 7th 1:24 pm);

– botnet C&C and direct data exfiltration (source: 192.168.120.7, DNS: 192.168.140.1, UTC time: from May 7th 2:30 pm to May 7th 3:02 pm).

1.3.2.2. Details of the method

The purpose of the method is to find one two-dimensional embedding per IP address per hour. This enables us to analyze the situation of the network and its temporal evolution. We aim to detect events and link them to the attacks by analyzing:

– clusters of IP addresses;

– trajectories of IP addresses.

Out method combines the node2vec and T-SNE algorithms. We apply node2vec to determine 128 (or 64)-dimensional embeddings. We then project these embeddings in a two-dimensional space for visual analysis. The node2vec algorithm uses random walks to determine each node's embedding. The choice of the parameters has to be consistent with the types of cyberattacks we aim to detect. For instance, when detecting "botnet C&C and data exfiltration through the DNS", the walks:

– are weighted by the number of bytes exchanged between nodes;

– have a maximum length of 4 (corresponding to the types of exfiltration patterns and rebounds that we are looking for).

In order to benchmark our method, we defined a baseline representation technique: 1 h of an IP address activity is encoded in a two-dimensional space using the following coordinates:

– x-axis: number of source messages (in degree) – number of destination messages (out degree);

– y-axis: two possibilities as follows:

- number of ports linked to source messages – number of ports linked to destination messages (with logarithmic transformation),

- amount of data out of the network through port 53 (DNS port).

This representation allows us to create IP clusters and determine receiving/emitting IP addresses. We refer to it as our "base method" in the remaining parts of the chapter.

1.3.3. *Visual results and interpretation*

1.3.3.1. *Network discovery attacks*

Figure 1.7 shows the results for network discovery attacks occurring on April 15 and April 16 for the base method: in the top left picture, IP 192.168.120.7 and IP 192.168.140.109 have very extreme coordinates. The attacking IP is at the bottom left of this picture (IP addresses sending a lot of messages and using a lot of source ports). The victim IP is at the top right (IP addresses receiving a lot of messages and using a lot of destination ports). The overall coordinate structure is consistent across the four pictures, making it possible to analyze IP trajectories over time. Nonetheless, we would like to see 192.168.120.7 and 192.168.140.109 very closely as they are highly connected during the attack.

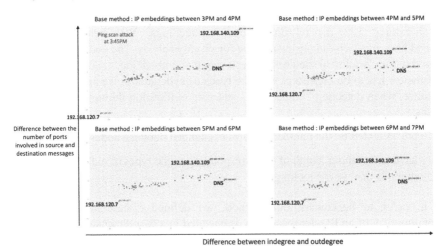

Figure 1.15. *Base method representations during a network discovery attack. For a color version of this figure, see www.iste.co.uk/briffaut/complexities1.zip*

Figure 1.16 shows the results for our embedding method (the number of ports involved in communications for each IP is proportional to the size of the dot: emitting IPs in red and receiving IPs in green).

We can see the benefit of using consistent embeddings:

– attacking and victim IP addresses are very close to each other in the embedding space;

– IP addresses are clustered in "communication groups" (IP 192.168.120.1 is the most important dot in a cluster of 20 IP addresses – this pattern is time-consistent);

– we can follow the temporal evolution of the network;

– we have a clear visualization of the attack.

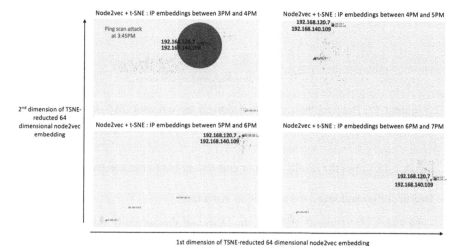

Figure 1.16. *Embedding-based representations during a network discovery attack. For a color version of this figure, see www.iste.co.uk/briffaut/complexities1.zip*

1.3.3.2. *Command and control attacks*

We also applied our method to more sophisticated attacks, botnet command and control attacks, in order to exfiltrate data through the DNS.

This was done by using port 53, traditionally dedicated to DNS communications.

Figure 1.9 shows the results for a DNS C&C attack occurring at 1:24 pm on May 7 for the base method: we analyzed the trajectory of the DNS (IP 192.168.140.1) over time: it tended to be in the top right corner (receiving IP on port 53, exchanging a lot of data) but suddenly mutated to the top left corner during the attack. This trajectory clearly exhibits a strange pattern.

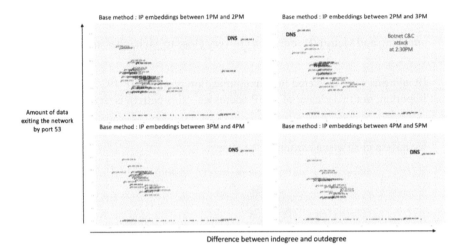

Figure 1.17. *Base method representation during a botnet C&C attack. For a color version of this figure, see www.iste.co.uk/briffaut/complexities1.zip*

Figure 1.18 shows the results with our embedding-based method (walk length = 4).

Due to our embeddings, we are able to visualize:

– the attack occurring at 2:30 pm: the DNS and the public IP addresses become very close with a very important weight, and their "attraction power" suddenly gets really high as a lot of emitting IP addresses mostly belonging to the 192.168.120.xxx and 192.168.150.xxx subnets get very close to those two IPs – the IP addresses attracted by the DNS are the "bots" that are being controlled and used by the attacker to exfiltrate data through the DNS;

– the position of the private IP addresses and the public IP addresses (outside the network);

– the groups of IP addresses communicating on port 53;

– the temporal evolution of the network and the trajectories of the IP addresses over time.

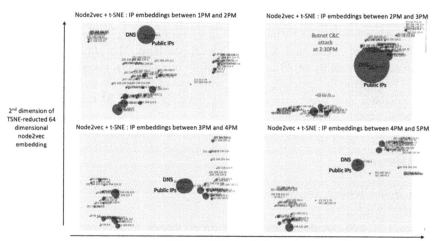

Figure 1.18. *Embedding-based representations during a botnet C&C attack.*
For a color version of this figure, see www.iste.co.uk/briffaut/complexities1.zip

1.4. Conclusion and future work

In this chapter, we presented a comprehensive study of complexity analysis in cybersecurity. We detailed the different types of virus and malware that companies are facing today and the complexity linked to the overall cyberdefense process. We showed that cyber-physical systems generate multivariate event data and can be characterized as complex and adaptative. As a result, two different types of anomalies (point-in-time anomalies and temporal correlation anomalies) need to be raised in order to detect multi-stage advanced persistent threats. In the last section, we presented UEBA machine learning as a way of reducing the complexity of a cyber-physical system. Our novel embedding-based framework has been designed to facilitate behavioral analysis, detect anomalies and link them to specific stages of APTs in order to facilitate SOC analysts' work. This framework is composed of machine learning algorithms, and it projects the representation of a network's assets in a visual two-dimensional space. It was validated with two types of attacks:

– network discovery through port scans;

– botnet C&C and data exfiltration through port 53.

This work was designed to enrich the toolbox of analysts working in security operational centers. The integration of machine learning (ML) methods into industrial processes gives hope for new growth drivers, in particular in the context of cybersecurity. The operational benefit of using machine learning methods is recognized but is hampered by the lack of understanding of their mechanisms, which is at the origin of operational, legal and ethical problems. This strongly affects the operational acceptability of AI tools. The ability of engineers, decision-makers and users to understand the meaning and the properties of the results produced by these tools will be a key success factor for the development of AI-enhanced security operational centers future years.

1.5. References

Adams, N. and Heard, N. (2016). Dynamic networks and cyber-security. *Security Science and Technology*, 1.

Adleman, L. (1988). An abstract theory of computer viruses. In *Advances in Cryptology – CRYPTO '88 Proceedings*, Goldwasser, S. (ed.). LNCS, vol. 403, Springer, Berlin.

Böhme, R., Laube, S., Riek, M. (2018). A fundamental approach to cyber-risk analysis. Report, Universität Innsbruck, Innsbruck.

Bonfante, G., Kaczmarek, M., Marion, J.-Y. (2006). On abstract computer virology from a recursion theoretic perspective. *Journal in Computer Virology*, 1(3/4).

Bradfield, J.C. and Stirling, C. (2007). Modal Mu-Calculi. In *Handbook of Modal Logic*, Blackburn, P., van Bentham, J.F.A.K., Wolter, F. (eds). Elsevier, Oxford.

BuzzFeedVideo (2018). You won't believe what Obama says in this video! [Online]. Available at: https://www.youtube.com/watch?v=cQ54GDm1eL0&feature=youtu.be.

Chawla, A., Lee, B., Fallon, S., Jacob, P. (2018). Host-based intrusion detection system with combined CNN/RNN model. In *IWAISe 2018 – ECML PKDD Conference*, Dublin.

Cohen, F. (1987). Computer viruses: Theory and experiments. *Journal of Computers and Security*, 6.

Filiol, E. (2004). *Les virus informatiques : théorie, pratique et applications*. Springer, Berlin.

Filiol, E., Helenius, M., Zanero, S. (2006). Open problems in computer virology. *Journal in Computer Virology*, 1(3/4).

Gartner (2015). Market guide for user and entity behavior analytics [Online]. Available at: https://www.gartner.com/en/documents/3134524/market-guide-for-user-and-entity-behavior-analytics.

Goodfellow, I.J., Shlens, J., Szegedy, C. (2014). Explaining and harnessing adversarial examples [Online]. Available at: https://arxiv.org/abs/1412.6572.

Jang-Jaccard, J. and Nepal, S. (2014). A survey of emerging threats in cybersecurity. *Journal of Computer and System Sciences*, 80(5).

Kirat, D., Jang, J., Stoecklin, M.P. (2018). DeepLocker concealing targeted attacks with AI locksmithing 2018. Report, IBM [Online]. Available at: https://i.blackhat.com/us-18/Thu-August-9/us-18-Kirat-DeepLocker-Concealing-Targeted-Attacks-with-AI-Locksmithing.pdf.

Kramer, S. and Bradfield, J.C. (2010). A general definition of malware. *Journal in Computer Virology*, 6, 105–114 [Online]. Available at: https://doi.org/10.1007/s11416-009-0137-1.

Mikolov, T., Sutskever, I., Chen, K., Corrado, G., Dean, J. (2016). Distributed representations of words and phrases and their compositionality [Online]. Available at: https://arxiv.org/abs/1310.4546.

Mirsky, Y., Doitshman, T., Elovici, Y., Asaf Shabtai, A. (2018). Kitsune: An ensemble of autoencoders for online network intrusion detection. In *Network and Distributed Systems Security Symposium (NDSS)*, San Diego.

Nesreen, K., Rossi, A., Rossi, R., Boaz, J., Willke, L., Willke, T.L., Zhou, R., Kong, X., Eldardiry, H. (2018). Learning role-based graph embeddings [Online]. Available at: https://arxiv.org/abs/1802.02896.

Or-Meir, O., Nissim, N., Elovici, Y., Rokach, L. (2019). Dynamic malware analysis in the modern era – A state of the art survey. *ACM Computing Surveys*, 52(5), article 88, 48.

Pingle, A. (2019). RelExt: Relation extraction using deep learning approaches for cybersecurity knowledge graph improvement. Cornell University [Online]. Available at: https://arxiv.org/abs/1905.02497.

SAIFE (2017). A new paradigm for securing today's porous perimeter [Online]. Available at: https://www.saife.io/wp-content/uploads/2017/08/SAIFE-WhitePaper-A-New-Paradigm-for-Securing-Todays-Porous-Perimeter-072317.pdf.

Sweet, C.R. (2019). Synthesizing cyber intrusion alerts using generative adversarial networks. Doctoral thesis, Rochester Institute of Technology, Rochester.

Viveros, C. (2016). Analysis of the cyber attacks against ADS-B perspective of aviation experts [Online]. Available at: https://pdfs.semanticscholar.org/e159/94e923c734e8c7a914b86ca25f1d7d9f14ea.pdf.

Yuan, Q. and Wei, S. (2018). Aligning network traffic for serial consistency and anomalies with a customized LSTM model. In *2018 IEEE International Conference on Progress in Informatics and Computing (PIC)*, Suzhou.

Zhu, L., Guo, D., Yin, J., Ver Steeg, G., Galstyan, A. (2016). Scalable temporal latent space inference for link prediction in dynamic networks. *IEEE Transactions on Knowledge and Data Engineering*, 28(10).

2

Complexity and Biology: When Historical Perspectives Intersect with Epistemological Analyses

The term "complexity" refers to the behavior of a system whose components interact with each other in a way that is difficult to predict. Beyond the reductionist breakthroughs that mark the history of biology, can we describe living phenomena as complex? When Hooke, in the 17th century, popularized the use of the microscope and highlighted the common structure of plants and animals, in so doing, did he reveal a complex world?

The subject of complexity in biology can be considered according to two themes, for which we will try to provide some answers. Indeed, complexity, from an epistemological point of view, refers, on the one hand, to the historical construction of the way in which we look at living beings and, on the other hand, from a scientific point of view, to the fact that our knowledge of the internal mechanics of living beings can be better described.

Therefore, our way of understanding biology could be conceived as a reading grid that remains dependent on a discourse that, for the sake of scientificity, seeks to reject outside its explanatory principles the arguments of teleology. This approach would be likely to lock biology into aporias marked by the eviction of concepts useful to the understanding of the world.

However, reducing biological complexity to the limits of our understanding does not permit us to understand life as the result of multiple processes that go from the microscopic scale to the subject taken as a macrocosm. Life is as complex as the set

Chapter written by Céline CHERICI.

of physicochemical phenomena at the basis of vital organizations. From this point of view, biology cannot be reduced to a constructivist conception, according to which all our knowledge would depend on the way in which our understanding formalizes it. Moreover, the knowledge we have is not neutral. On the contrary, it is impregnated with representations of great historical profoundness, which sometimes seem incompatible with each other. Therefore, we approach conceptual couples, such as invariance and variability and chance and necessity.

Within living systems, complexity as a phenomenon seems to emerge in stages. Therefore, to speak of complexity implies evoking a functioning by interaction between scales of life or a biology conceived in terms of *Russian nesting dolls* or nested systems. Two major principles seem to intervene in a repetitive manner: the accumulation of identically replicated elements and then their arrangement in more complex sets of which these elements form parts. Chapouthier (2018) proposed to qualify these sets with the term *mosaic*:

> In a mosaic, in the artistic sense of the term, the whole, the "totality" represented, leaves a certain autonomy of appearance to the tesserae that compose it and which retain their form, their color or their brilliance, in the same way that an organ leaves a certain autonomy of function to the cells that compose it, as with the organs that constitute an organism, an animal society to the individuals that form it, etc. (p. 4, author's translation)

This artistic metaphor makes it possible to highlight a dual approach necessary to the biological sciences toward the microscopic and the distanciation needed for the understanding of the macroscopic. Indeed, if getting closer to a mosaic allows us to observe and describe the units, the tesserae, moving away from it gives us an overall view that would be lost by the meticulous description of the parts. Both at the present scientific level and throughout history, such a ballet seems to take place, through which scientists either approach the smallest components of the living world or seek a perspective closer to the macroscopic scale.

Therefore, we provide philosophical elements to reflect on the problem of complexity through the lenses of biology, epistemology and the history of biology.

In the context of the constitution of molecular biology, the following question can be asked: would it be possible for modern biology, autonomous as a science of the living, to continually escape technical and physicochemical reductionism? Does not living matter exceed, in the knowledge that we can have of it as a whole, the limits of studies on the mechanisms that govern its parts?

First, we must recall some of the stages of biological complexity from the history of life sciences in order to highlight the interdependence of the way we look at it with the representations at work during this process of knowledge acquisition; then, in a second step, we explore the incessant game that life plays between potentialities and actualizations in order to understand to what extent biological knowledge cannot do without the concepts of invariance and potential. Finally, we examine the links between theoretical biology and contemporary biotechnologies, asking whether the relationships that bind them are not situated at the heart of a dichotomy between experimental techniques that work on the reduction of biological phenomena and a theoretical biology whose terms and principles never cease to overload them.

2.1. Complexity throughout the history of thought on living

2.1.1. *The roots of thinking on complexity*

From the teleological and integrative approach that Aristotle proposes on the formation of life, an idea emerges according to which the progressive development of organized structures is due to the phenomena of complexification that have occurred in a form that has the ability to repeat itself while acquiring new parts and then new functions.

If, already in Aristotle, we find the representation of a continuous interweaving of different levels of the functional complexity of the living through the interplay of plant, animal and rational souls, then we can consider that a conceptual weaving between form and function, integration and expression, but also an emergent understanding of the different forms progressively taken by the living, has been in place since antiquity. Aristotle appears as the thinker of a biology that shows a concern for an integrative explanation of life as a functional whole. This movement of reading nature implies numerous difficulties concerning the identification, classification and apprehension of the convergences and divergences that exist between species:

> However, in the hierarchy of beings, the progression is made by imperceptible gaps. Among these forms that slide one over the other, it is very difficult to decide where each domain begins and ends. A sponge, who will say if it is a plant or an animal? And a coral, is it really a stone? (Jacob 1976, p. 31, author's translation)

In Aristotle (2009), we find two types of classifications – one by groups divided into subgroups, and a second more integrative and based on the concept of plural souls, where the most complex level encompasses the previous ones:

> The soul is the cause or source of the living body. The terms cause and source have many senses. But the soul is the cause of its body [...]. It is (a) the source or origin of movement, it is (b) the end, it is (c) the essence of the whole living body. That it is the last, is clear; for in everything the essence is identical with the ground of its being, and here, in the case of living things, their being is to live, and of their being and their living the soul in them is the cause or source.

This entanglement of forms is visible until the 16th century, when possibilities directed by God, the soul or the cosmos still predominate. In the 17th century, a scientific project involving the deciphering and reading of nature by the human understanding was developed, proposing a reduction in its foundations and putting in continuous correlation the worlds of the inorganic and the organic, the second being founded on the first:

> Philosophy is written in this grand book, the universe, which stands continually open to our [p. 238] gaze. But the book cannot be understood unless one first learns to comprehend the language and read the letters in which it is composed. It is written in the language of mathematics, and its characters are triangles, circles, and other geometric figures without which it is humanly impossible to understand a single word of it; without these, one wanders about in a dark labyrinth. (Galilei 1968, p. 232)

The explanatory power determines the value of the reductionist hypothesis. Indeed, on the condition that the complexity of nature, notably visible through its polymorphism, is reduced to the simplicity of the elements that compose it, it is a question of trying to understand it while giving leave to a universe marked by finalist discourses. It is thus not a question of possessing knowledge of all things, but of choosing signifiers likely to become consensual and which permit the internal exploration of phenomena. Until the 18th century, the concept of the *living* is rooted in the inanimate by virtue of the physicochemical unity of the laws of reading the world. Therefore, the reductionist reading grid, based on the resolution of the apparent complexity by the simplicity that underlies it, becomes a source of new knowledge.

In order to shed light on a mechanics that seems inaccessible outside of a finalist way of thinking, and in order to reach an objective truth, Descartes established as a methodological principle the fact of dissociating the object not only from the subject that observes it, but also from a natural tendency to conceive of the living through the prism of a pre-established harmony of the world. By reducing complex mechanisms, sets of organs or complete organisms to the elements and parts that compose them, the philosopher permitted the scientific eye to seek a meaning to each micro-part of a whole. Therefore, in order to understand its polymorphism, it is necessary to cut nature into small elements, separated from each other, reduced and simplified. This approach is the basis of a theory of scales, which is completed by the microscope, within which biological structures develop and can be studied in isolation through the prism of discontinuity. However, at the end of this reduction of the complex to the simple, the study of living beings found itself faced with a contradictory epistemological and scientific situation, where invariance and random mutation had to be articulated in order to understand the links between the whole and the parts. From this perspective, complexity arises simultaneously from crossed historical views and the difficult scientific reading of biological phenomena. The necessary articulation of the microscopic scales of the physical and chemical elements that make up living beings to the macroscopic dimensions of organic ensembles that function as a whole has tended to generate an opposition between scientific reductionism and an overall view that restores the autonomy of living beings. On the one hand, there is the mechanistic interpretation of the organism and, on the other hand, the obvious meaning of certain phenomena, such as embryological development.

2.1.2. *From machinules to cells: An ordered complexity?*

Based on the idea of the repetition of organisms, the notion of species, at the end of the 17th century, aimed at understanding the formation of living beings through the prism of the category, the group and the whole. It required the exploration of the internal embryo-logical mechanics and found at the macroscopic level the notion of the continuous generation of species. The visible permanence of species was then confronted with variations notably brought to light by comparative anatomy. To find the natural order in the development of living beings was one of the aims of natural history. While Linnaeus (1735) conceived a taxonomy based on the morphological and anatomical criteria of each organism, categorizing living things by classes, orders, families, groups and species, Jussieu (1778, pp. 175–197) developed a new system of plant classification:

> It is therefore necessary to appeal to an external element that is not based solely on the visible structure of beings but on the permanence of this structure through generations. The concept of species was thus born, at the end of the 17th century, from the need for naturalists to give their classification a point of support in the reality of nature. (Jacob 1976, p. 61, author's translation)

From 1665[1] onwards, the microscope made it possible to capture the phenomena of generation, through the notions of forms and movements, and highlighted the

---

1 In 1665, Robert Hooke (1635–1703) observed what he soon came to consider as the constituent units of cork: dead cells which made him think, by their appearance, of the cells of a

presence of eggs in females and animalcules in males; even if without recourse to formative vital impulses, the internal mechanics remained very mysterious. This point partly explains the persistence of the pre-formationist model that guarantees the harmonious repetition of the world. Therefore, as early as the 17th century, scientists set out in search of a unit that could be considered as an infinitely small machine having its own functions and forming all living beings. This dynamomechanical conception of the organism made it possible to resolve the question of its genesis by studying the mechanisms of these micro-units and the way in which they aggregate, thus sketching the cellular model. It then appears that beings that are different on a macroscopic scale possess the same internal organization under the microscope. A biological unit is thus introduced in the living beyond its apparent variety:

> Malpighi apprehends living bodies as extremely complex arrangements of "little machines", *machinulae*, which the microscope and analogies with simpler living beings bring to light. His anatomy promotes a decentralized view of the organism and points to the uniformity of composition between the whole, the parts, and the parts of parts. (Andrault 2012, p. 297, author's translation)

It appears that the history of complexity in biology is constituted in as many levels as those that permit us to understand this discipline, both in its singularity and autonomy and in the simplicity of its elements. Therefore, from the 17th century to the experimental physiology of the 19th century, a *hiatus* emerged between the vital phenomena described on the scale of their visible manifestations and the knowledge of the elements that compose them, whether formal, such as the cell, or chemical or physical. Life, although rooted in physics, overflows the latter through a multitude of ways of creating living forms that escape prediction:

> the vital force directs phenomena that it does not produce; the physical agents produce phenomena that they do not direct. (Bernard 2022, p. 51)

Mendel (1907, pp. 371–419), by giving a mathematical language to the study of the transmission of traits in botany, contributed to a movement of dissolution of an organic whole into significant letters, signs or simple elements and turned the representation of an animated and reducible organic world into an alphabet accessible to instruments as well as to reason. However, it seems that this reductionist program has been overwhelmed by the variability of compound bodies.

---

monastery. This is why he retained the name. Then he observed what he identified as identical cells within the plants. Although he used the word "cell" in a more morphological than functional sense, he lay the epistemological foundations of cell theory. Hooke described these small structural units in his work entitled *Micrographia* (1665).

We can consider that biology has taken two directions, from the 19th century onwards, which intertwine, but which are also opposed in the perspectives they cross, microbiology and macroevolution: on the one hand, from 1800 and the developments of transformist thought, biology, in a macroscopic dimension, has been interested in the biological object in its environment and in its group. This dimension can be compared to a taxonomic approach aiming at classifying groups according to their common attributes, and thus at understanding their relationship links. On the other hand, from a microscopic perspective, the biological sciences become interested in a reductionist focus, in which a whole is reducible to its parts.

Complexity lies in the necessary articulation of these two dimensions. One of the scientific and epistemological questions remains: are the functions as the result of a set of phenomena reducible to the description of the parts from which they seem to emerge?

> Without doubt, the whole can have properties of which the constituents are devoid. But these properties result from the very structure of these constituents and their arrangement. (Jacob 1976, p. 15, author's translation)

The shift from natural history to biology reflects a new approach that is no longer solely interested in morphological differences but also in the similarity of properties that distinguish living bodies from inert bodies:

> All that is generally common to plants and animals, as well as all the faculties which are proper to each of these beings without exception, must constitute the unique and vast object of a particular science which is not yet founded, which does not even have a name, and to which I shall give the name of biology. (Lamarck 1815, vol. 1, p. 49, author's translation)

2.1.3. *The organism: An autonomous complexity*

This continuity between the animate and the inanimate encountered division at the end of the 18th century, initiated by Bichat (1771–1802), a physiologist who exposed the existence of a vital principle, characteristic of a life intrinsic to organic matter and according to which "Life consists in the sum of the functions, by which death is resisted" 1799–1800, p. 1). Therefore, he developed the idea of a particular organization at the foundations of life. The human and animal machine reveals itself to be alive insofar as its laws and its characteristics are proper to the organization of matter. Even static electricity becomes animal within electrophysiology, thus making the living electrocentric. From the discovery of animal electricity by Galvani (1953), who spoke of "neuroelectric fluid" (p. 64), electricity became internalized, entering into the functioning of the deep tissues. Therefore, the expressions animal machine, human machine or human animal machine punctuate certain 19th century treatises[2]. It is more a question of underlining the regular functioning of the living than of reducing it to a machine:

> If these different fields of physiology can thus be analyzed given that they have become accessible to the methods and concepts of physics and chemistry, in return the analogies observed and the models used contribute to a radical transformation of the representation that the end of the 18th century has of living beings. (Jacob 1976, p. 53, author's translation)

Between the 18th and 19th centuries, the notion of the organism emerged from the framework of a living being that was organized as a set of emerging functions and no longer as the mechanical articulation of simple elements. If there is always an organization to be discovered behind the whole, it seems to be divided according to two scales: those of the components, of the parts that form the organs, and that of the animal economy. Therefore, the microscopic explorations of the animate, which have been developing since the last third of the 17th century, are coupled with a natural history that becomes biological and which, by classifying species, seeks to objectify elements common to the different forms of life in order to guarantee an apparent continuity in nature that goes beyond its polymorphism. Taxonomy contributes to the weaving of a macroscopic reading grid that allows each biological individual to be reinscribed in the group and the environment. If it is a question of finding lines of convergence and divergence to explain living beings, this approach requires the association of two concepts: teleonomy and chance. Teleonomy replaces a finalist discourse of a philosophic type to elaborate the idea of a direction at the heart of living beings that guarantees their order, whereas chance progressively forges an empirical and singular reality through the multiplicity of natural productions.

---

2 For example, these expressions occur more than 25 times in 1804 in Giovanni Aldini's work. See Aldini, G. (1804). *Essai théorique et expérimental sur le galvanisme, avec une série d'expériences faites en présence des commissaires de l'Institut national de France, et en divers amphithéâtres de Londres*. Piranesi, Lucchesini, Paris; Cherici, C. (2020). *From Clouds to the Brain: The Movement of Electricity in Medical Science*. ISTE Ltd, London, and Wiley, New York.

2.1.4. *The emergence of complexity between comparative anatomy and embryology*

Developments in biology should not be thought of separately from the history of comparative anatomy, taxonomy or embryology. Indeed, as soon as one has to compare several species or different biological scales, such as those of the molecule or the cell, the biologist is led to call upon general notions defined over the past centuries within the framework of anatomical comparison. Homology designates a field which determines the specific characters of several species and attributes them to a common ancestor. This notion is deeply linked to comparative anatomy, to classification disciplines, and is connected to evolutionary sciences, insofar as the anatomical structures described share an evolutionary history. Its history cannot therefore be reduced to that of 20th century biology.

Therefore, the comparison of animal forms found itself very early on at the center of a web of relationships between different fields of the life sciences in the process of specialization. The development of biology has gone through a movement of internalization of a complex network of studies, first of forms and structures, then of microelements of the living. If this movement is technical and has involved the invention and improvement of the microscope, it is also theoretical and involves the construction of different reading grids in which the polymorphism of life and an anatomical, physiological and functional unity are understood correlatively. From the reduction of form to its parts to the concept of organism, conceptual mutations are articulated that accompany the development of techniques:

> This time, it is a question of the progress of the idea of organism, conceived as an indivisible whole, in relation to a critique of strict mechanism and the return in force of vitalism as well as to the reflection, undertaken notably by Kant, on the role of finalism in the natural sciences. (Schmitt 2006, p. 152, author's translation)

Between the end of the 18th century and the beginning of the 19th century, the concept of organism was defined as an organized whole that interacts with other vital organizations. Therefore, according to Kant's (1965) analysis of Descartes' mechanistic metaphors:

> In a watch, one part is the instrument that makes the others move, but one gear is not the efficient cause that produces another gear; certainly a part exists for another, but it is not by this other part that it exists [...] Hence an organized being is not a mere machine. For a machine has only motive force. But an organized being has within it a formative force, and a formative force that this being imparts to the kinds of matter that lack it (thereby organizing them). This force is therefore a

formative force that propagates itself – a force that a mere ability [of one thing] to move [another] (i.e., mechanism) cannot explain. (p. 193)

This long quotation highlights the break of vitalist thought with Cartesian reductionism, the link with the emergence of the vital organizations which exceed the knowledge of the parts, but also returns us to the permanent paradox of a living machine which requires mechanical phenomena, the arrangement of simple elements, while being the theater of forces specific to the living.

In the early 19th century, systems theory, developed by the Italian neuro-anatomist Vincenzo Malacarne[3], accounted for the organism as a hierarchical set of vital functions having the relationships between organs as their support (Cherici 2016, p. 12). Physiology adds to anatomy the step wherein biological analysis is based on the relationships between organs and groups of organs on the interaction between parts whose understanding is articulated to functions.

This is how Malacarne develops his research on the reciprocal influences that the different systems of the human organism have on each other. He seeks to conceive the functioning of the body in correlative and integrative terms of organs and physiology. From the tissue system, described as universal, to the smallest systems of the organism, the body is divided into as many sets in order to ensure its functions:

> The installation of the concept of organization at the heart of the living world has several consequences. The first is that of the totality of the organism, which now appears as an integrated set of functions, and therefore of organs. What must be considered in a being is never each of the parts taken in particular, but the whole [...]. (Jacob 1976, p. 99, author's translation)

The notion of bodies organized according to their constituent parts precedes that of the organism; and from the first to the second, the particles come to life, becoming, for example, organic molecules in Buffon (1888, p. 203). So many elements internal to matter, conceived beyond the level of atoms, describe the living particles in movement in bodies. As early as 1714, the Leibnizian model of the monad made it possible to conceive, by analogy with the ingredients of life observed

---

3 *Della esistenza di molti sistemi e della influenza loro nella economia animale* (Malacarrne 1803). The French version received a medal in 1803 from the learned society created by Bichat, the *Société médicale d'émulation de Paris*, of which Malacarne had been a member since 1797.

under the microscope, of living particles which, during embryological development, realize a vital organization, giving the idea that matter is endowed with memory. The thesis of monadology presents this imbrication of matter and life in an infinite process of arrangement:

> From this we see that there is a world of created things, of living beings, of animals, of entelechies, of souls, in the smallest particle of matter. Every portion of matter may be conceived as like a garden full of plants and like a pond full of fish. But every branch of a plant, every member of an animal, and every drop of the fluids within it, is also such a garden or such a pond. [...] There is, therefore, nothing uncultivated, or sterile or dead in the universe, no chaos, no confusion, except in appearance. (Leibniz 1846, p. 474)

While Bichat's vitalism advocates the indivisibility of organic matter, the developments of cellular theory sequenced the properties of life present within each cell:

> What is important here is not so much that cells are found in all tissues or even that all organisms are made up of cells, but that the cell itself possesses all the attributes of the living, that it represents the necessary source of every organized body. (Jacob 1976, p. 132, author's translation)

The study of living beings as the result of vital complexifications of matter is reinforced by Lamarck's transformist thinking, which does not submit these biological transformations to an external will, but crosses their internal modifications with a pressure from the environment in a macroscopic perspective. Complexity lies in the biological understanding of life forms, in the scientific project of understanding them in their elements, their parts, but also in spatial and temporal correlations with each other. The development of biology at the beginning of the 19th century, claimed by Lamarck, far from eliminating a teleonomic thought, imprints to the organic matter, its movement, its tendency to transformation and its self-realization:

> The idea is certainly not a metaphysical object, as many people like to believe; it is, on the contrary, an organic and consequently quite physical phenomenon, resulting from relations between various matters, and from movements which are executed in these relations. (Lamarck 1830, p. 292, author's translation)

The stability of species over an extremely long period of time, associated with a permanence of chemical structures, has been emphasized by an intellectual construction guided by teleonomy. However, the latter, by giving a hand to natural selection, has also promoted a variety, a contingent selection of nature's possibilities:

> The fact that, in the evolution of certain groups, we observe a general tendency, sustained for millions of years, to the apparently oriented development of certain organs, testifies to the fact that the initial choice of a certain type of behavior (in the face of a predator's aggression, for example) commits the species to the continuous perfection of structures and performances that are the support of this behavior. (Monod 2014, p. 164, author's translation)

From Lamarck to Darwin, a living being is conceived of as a set of often infinitesimal biological movements interacting with the development of new organs and/or new functions. When Darwin brings into play the role of a contingent and blind natural selection, he puts an end to the idea of divine creation and highlights the biological tendency to variability of organic matter:

> In the long term, it integrates mutations; it arranges them into adaptively coherent sets, adjusted over millions of years and millions of generations, in response to the challenge of the environment. It is natural selection that gives direction to change, that guides chance, that slowly, gradually, elaborates more and more complex structures, new organs, new species. (Jacob 1976, p. 33, author's translation)

Darwin introduced the idea that life is both permanent and variable. Our biological world is therefore only one of the possible worlds, since life creates a space of possibilities based on interactions between different levels of combinations and arrangements of biological units, tissues or organs:

> We live in a myriad of emergent becomings that is "impreconceivable", literally unimaginable. Since we cannot write any law describing the particular emergence we experience, although our foundations are based on physics, we are beyond it. The living world is not a machine. (Kauffman 2019, p. 162, author's translation)

Introducing contingency at the level of biological variability, Darwin links it to a teleonomic perspective on generation. Indeed, the pangenesis he describes (Darwin 2008) is part of both the study of microstructures animated by themselves, subject to the laws of evolution as well as to the action of natural selection, to the repetition of each type of living thing, but also to biological variation, insofar as each fragment of the body produces a small germ of itself, or "gemmule", which develops with a view to reproducing that fragment in the following generation.

While, with Mendel (1907), biology acquires a form of mathematical language as a condition for the production of knowledge based on measurement, with Darwin it gains organic depth, as matter becomes a field of action for time and space, variation and permanence.

The birth of molecular biology, from the meeting of biochemistry and genetics around 1940 (Morange 1994, p. 17), represents a new stage in scientific reductionism, insofar as this discipline seeks to interpret the properties of the organism through the structure of the molecules that constitute it, thus penetrating to the heart of the microscopic universe. The metaphors with the machine, which include the notion of program, go in the direction of a reading of the living being in connection with causal mechanistic principles and support the development of functions. However, at the heart of these constructed representations, the dimensions of the continuous and the discontinuous, the visible and the invisible (cell, molecule), stability and instability do not cease to overflow the reduction of the whole to its parts and to give again its place to the complexity of elements in constant movements. This perspective is born from a historical movement composed of Cartesian mechanism, vitalist thinking and the evolutionary and experimental biology representative of the 19th century.

In the 20th century, genetics was marked by the relationship between the phenotype and the genotype, which it only partially expresses. The cell, with its moving nucleus, represents a universe with many properties, and these studies cross the internal variability of species, itself woven into their stability – the notion of transmission is redefined from its internal mechanisms to its external effects on the individual and the group:

> In fact, in less than twenty years, cellular theory in its final form appears, the theory of evolution, the chemical analysis of the great functions, the study of heredity, that of fermentations, the total synthesis of the first organic compounds. (Jacob 1976, p. 196, author's translation)

Organization becomes the goal of biological studies, since it refers to the interactions between the level of complexity of the microscopic scale with the macroscopic scale, which includes the notions of group and environment. The complexity of bodies increases as the techniques of exploration of the living world progress. On the one hand, the tools of investigation go further and further back into organic structures, going as far as the components of DNA (Crick and Watson 1953); on the other hand, if biotechnologies remain in a reductionist perspective, insofar as they must reduce the elements to synthesize and replicate them, the living in their multiplicity do not allow themselves to be locked in.

From bacteria to human beings, there is a horizon made up of the movements from the units of the chemical machinery, from which it is up to the scientist to find harmony at the level of structures and modes of functioning. It is the reading and reproduction of the text written in the DNA that has made it possible to objectify a universal harmony that ensures the invariance of species. However, this invariance evolves and modifies its forms at the rhythm of chemical and biological variations or because of external accidents. From the acronym CHNOPS (carbon, hydrogen, nitrogen, oxygen, phosphorus and sulfur), which refers to the molecular ingredients, the components of DNA, the idea of a permanent interaction between the physicochemical levels and the genetic mechanisms stands out.

If the cell can be considered as a closed universe, it evolves. Micro-modifications are not lacking, and under the effect of their accumulation, any structure of an organism is gradually altered, or even destroyed if the errors are too numerous. Biological chance, that of a contingent variability, is a source of novelty and has not been eliminated from scientific discourse. Counterintuitive, because it is unpredictable, it does not cease, by associating itself with the permanence of life, to make our intuition of the living more complex and the application of our paradigms and representations to its mechanics. When the accident is inscribed in the DNA, it is replicated. If its emergence is due to chance, its replication is a necessity.

Biology is a discipline involved in a historical construction that is the result of the developments of the many disciplines that precede and found it. These disciplines are impregnated with representations, among which are teleology, which became teleonomy, and chance, thus deepening the complexity of the discourse on life. Complexity in biology is a notion with compartments, resulting from historical constructions, as well as from the complexity of viewing living beings at different scales. Therefore, from the repetition of structures to variability, life oscillates between structural and functional potentialities and actualizations. However, what laws do these actualizations obey? From which paradigms do we interpret the links between invariance and contingency?

2.2. The living: Between potentialities and actualizations

2.2.1. *Teratology to better understand the links between actualization and potentiality in the living*

At the margins of the embryological normal and pathological, developments in teratology seem to be a good example to illustrate the infinite biological game between the possible and the actualized. Indeed, as long as the accident of development is attributed to supernatural or metaphysical causes, it cannot be objectified as a process necessary for the repetition of its category. To understand monstrosity through the prism of fetal development implies considering the fact that deformation obeys a plan, a possible biological model. Deformations and teratological variations are thus objectified by the change in status of the observation of *human monsters* in the 18th century. If anomalies obey rules of coordination and arrangement, then new analyses of embryology are made possible. Teratogeny sheds light on the latter as a domain of borderline cases where the *monster* is possible and its actualization not necessary. In the 19th century, anatomopathological preparations also illustrated the regularity of different teratological deformations. Teratology allows for the articulation of notions of structure, function and actualization in a given environment. Vincenzo Malacarne, between 1802 and 1811, proposed a classification of teratological cases whose anomalies are repeated according to a principle of arrested development:

> In fact, I have before my eyes four perfectly acephalic human monsters, that is to say, who are completely lacking the head, and who, what will seem the strangest, lack the main viscera of the thorax, which are the heart, the largest vessels, the lungs, the trachea, the esophagus, the thoracic duct, the liver, the spleen, the pancreas, the ventricle, as well as a large part of the intestines. (Malacarne 1811, p. 2, author's translation)

The term *perfectly* correlates with the idea that the monsters described by Malacarne are coherent from the point of view of the model to which their deformation obeys, which was therefore only one of the possibilities of their development. Malacarne emphasizes the activity of a constituent matter in acephalous fetuses whose morphology obeys a unique compositional plan, thus enunciating one of the foundations of modern teratology: the principle of arrest of organic development applied to comparative organology, as well as to teratogenics. From his studies is extracted a classification, where each category is given a particular name from the Greek, which indicates a primary morphological character: he classifies, for example, micromelia, which is the smallness of some limb, and describes observed cases of microcephaly.

In 1824, in his *Anatomie comparée du cerveau dans les quatre classes d'animaux vertébrés appliquées à la physiologie et à la pathologie du système nerveux* (Comparative Anatomy of the Brain in the Four Classes of Vertebrate Animals as Applied to the Physiology and Pathology of the Nervous System), Serres set out the idea that the natural classification of species is based on the morphology and progressive complexification of the encephalon. This conception was inherited from research on comparative anatomy. The principle of the arrest of organ development fits into this scale by making it possible to include deviant organisms:

> Secondly, if the organisms' march of developmental progress is halted, these organisms struck by an arrest of time will necessarily have to reproduce those of some animal from a lower rank than the one observed. (Serres 1859, vol. 1, p. 27, author's translation)

Therefore, development in its normal and pathological forms accounts for the complexity inherent in the persistence of species, the repetition of incomplete structures and the permanence of a teratological possibility. In the embryological perspective, the question of sexual reproduction appears as a condition of variability, the latter being taken in the complexity of a repetition always submitted to the actualization of what is not yet but that which could be. Sexuality can thus be conceived as an effect of the struggle for existence within the same species, but also as a double set of conditions of variation and repetition. The importance of sexual reproduction and the pressure it exerts on the animal group can already be found in Erasmus Darwin's *Zoonomia; or the Laws of Organic Life*)[4]:

> As they reproduce, organisms propagate the way they organize their functioning, the way all these parts fit and work together. Organs are part of this organization and exist both for and because of the whole. (Kauffman 2019, p. 15, author's translation)

From the moment that sexuality is a necessary condition for reproduction, each being is composed, insofar as it results from the reassortment of the two parent programs. However, in organisms where reproduction is not sexual, the genetic program is exactly copied. All the individuals that make up a generation are identical, except for a few rare mutants that represent the emergence of genetic variation. Such populations can only evolve if they are selected and/or under pressure from the external environment. The complexity lies in the continuous recombination of genetic programs through sexual reproduction, as well as in the link between adaptation, unpredictable mutations, and the selection of the latter. Therefore, sexuality can be presented as a criterion for accelerating evolution, a condition for the development of evolutionary advantages that permit the rapid juxtaposition of favorable mutations that would remain separate in identically replicating species. By interweaving the macroscopic scale, of the species or the individual, with the microscopic scale, biology has become the site of a paradox: the

---

4 Erasmus Darwin (2020) writes, for example, that "The final cause of this contest amongst the males seems to be, that the strongest and most active animal should propagate the species, which should thence become improved".

simplicity of the ingredients of life (carbon, hydrogen, nitrogen, oxygen, phosphorus and sulfur) is the basis for the complexity of their development, the latter being caught up in a structure where chance cannot dispense with a dimension that is necessary for the perpetuation of life.

If the history of biology is articulated around those of comparative anatomy, classification, mechanism, techniques and vitalism, it also bears their representations. By wanting to eliminate the finalist discourses from their explanations, has the scientist not become locked in an epistemological contradiction? The non-teleological reading of the laws of life cannot avoid a structural invariance. The status of teleonomy in the biological sciences is an illustration of this paradox. Enunciated by Colin S. Pittendrigh in 1958, it is differentiated from teleology in these terms:

> Biologists for a while were prepared to say a turtle came ashore and laid its eggs. These verbal scruples were intended as a rejection of teleology but were based on the mistaken view that the efficiency of final causes is necessarily implied by the simple description of an end-directed mechanism. [...] The biologists long-standing confusion would be removed if all end-directed systems were described by some other term, e.g., 'teleonomic,' in order to emphasize that recognition and description of end-directedness does not carry a commitment to Aristotelian teleology as an efficient causal principle. (p. 394)

Teleonomy remains in the representation of a living being that realizes organizations whose end is the expression of their functional structures. This concept is taken up by Jacques Monod who highlights the fact that simple molecular mechanisms, in their behavior and their development, can give the impression that living beings tend towards a goal.

Therefore, the living has a biological autonomy made visible by variability, but always linked to the notion of necessary realization or invariance. Matter has characteristics expressed by laws internal to its microscopic elements which depend on physics and chemistry and the knowledge of which seems insufficient to understand its polymorphic development:

> The organism is a machine that builds itself. Its macroscopic structure is not imposed on it by the intervention of external forces. It constitutes itself autonomously, thanks to internal constructive interactions. Although our knowledge of the mechanics of development is insufficient, we can nevertheless affirm that the constructive interactions are microscopic, molecular, and that the

molecules involved are essentially, if not solely, proteins. (Monod 2014, p. 68, author's translation)

The notions of organism and systems, which have been used since the end of the 18th century, have continued to develop in order to help characterize organic units whose functioning depends on movements and phenomena internal to the cellular system and grasp the idea that the metabolic activities of the cell depend on a physiological harmony, on a coherence of which we do not have all the keys to understanding. The description of the microstructures that constitute living organisms requires associating the ideas of hierarchy, permanence, variation and harmony. Therefore, the study of microscopic systems reveals that the complexity of living beings exceeds what the study of the functions of organisms conceived of as a *whole* has been able to suggest:

The analysis of allosteric interactions shows first of all that teleonomic performance is not the prerogative of complex, multicomponent systems, since a protein molecule is already shown to be capable not only of electively activating a reaction, but of regulating its activity according to various chemical information. (Monod 2014, p. 106, author's translation)

While biology is complex, it is because it is structured by an incessant and unpredictable play of potentialities and actualizations. We speak of biological revelation, insofar as the cell, the molecule or the elements of the DNA already contain the foundations of what will be the body to be realized, which does not exclude the variation or the pressure of the chemical and industrial environment during the stages of development. This last point puts into perspective the image of a cell closed in on itself. Therefore, the development is to be thought of in interactions with variables, such as temperature, which reinforce the unpredictability of future variations, which join the domain of probabilities. The latter are the many possibilities which are not necessary, but which potentially can be actualized.

Therefore, DNA was described as the result of its constituents: four complementary elements, the nucleotides, which are adenine, thymine, guanine and cytosine. Their formalization in A, T, G and C could transmit a representation according to which contemporary biology possessed an entirely readable alphabet of the living. This reduction of the DNA elements into letters does not make it possible to account for the contingent aspects of possible formations. If developments have the plan of future structures contained within the constituents themselves, the activities of the cells are caught in a simplistic dialectic at the level of the reading their structure and a complex one at the level of their role in the formation of a tissue or the specialization of an organ:

It is therefore possible that the "cognitive" properties of cells are not the direct manifestation of the discriminative faculties of some proteins, but only express these faculties in very roundabout ways. Nevertheless, the construction of a tissue or the differentiation of an organ, macroscopic phenomena, must be considered as the integrated result of multiple microscopic interactions [...]. (Monod 2014, p. 118, author's translation)

In the 1950s, a fundamental discovery was made in Watson and Crick's understanding of how DNA codes for proteins. Indeed, Sanger determined the complete amino acid sequence of two polypeptide chains of bovine insulin, A and B, thus describing the first complete sequence of a globular protein. By singling out these sequences, he moves away from the representation of an amorphous protein, showing that it instead has a defined composition (Sanger 1959, pp. 1340–1344).

The role of chance is put forward in the unfolding of a protein sequence less by admission of ignorance than to emphasize that it is very complex to formulate rules, theoretical or empirical, that would permit us to predict the nature of all the residues or portions of free amino acids. However, necessity is at the very heart of these protein sequences and ensures their repetition at the level of a population of molecules. One can consider that there exists, within the microstructures of the living, an invariance of structures implemented by contingency "[...] captured, preserved, reproduced by the machinery of invariance and thus converted into order, rule, necessity" (Monod 2014, p. 128, author's translation).

In understanding how virtuality and chance, natural selection and adaptation to the environment are intertwined, embryology makes it possible to highlight the links between what can possibly happen and the element of chance in this becoming the reality, as well as the unpredictability of the effects of the environment on the embryo. Therefore, the theories of evolution, by giving a temporal and geographical dimension to life, have made it possible to think of change as necessarily due to chance based on the hypothesis according to which the variations in the properties of organized bodies are based on the permanence of a structure housed in the cell that is not necessarily actualized:

However, in addition to natural selection, we now know of a whole series of mechanisms involved in evolution: for example, genetic drift, the random fixation of genes, the indirect selection brought about by a linkage of genes, the differential growth of organs, etc. (Jacob 1970, p. 43, author's translation)

2.2.2. *Time, a key concept for understanding the interactions between the possible and the actualized*

Time plays a fundamental role in the development of biology and has made it possible to set up a double reading grid: that of the temporality of ontogenesis and that of the history of life on earth, the second influencing the first. This double temporal reading of biology is integrative: not only because, through the perspective of an Aristotelian heritage, it permits us to study the unfolding, during the formation of a being, of an interlocking biology in which complex mechanisms include the simplest ones; but also because it allows us, by understanding the relations between animal and plant species, to think of these relations over the course of history. Heredity can thus be read over a long period of time, but also through individual time, and is based on three principles: function, mutation and reassortment. It is the gene that carries these perspectives of potentiation and actualization because it represents the unity of function, mutation and recombination. Let us take the example of cytochrome c to illustrate the importance of time in understanding the development of species. It is a small molecule present in the cell membrane of prokaryotes and in the inner membrane of the mitochondria in eukaryotes. It links the physical and chemical levels of matter and functions as an electron shuttle, in addition to participating in respiration. Its molecular history has shown that after a threshold of unintended variation over time, structures differ in various species. It is considered to have existed since the beginning of life on earth and has been found in most living things, from plants to animals to bacteria. As a place of variability and invariance, it is extremely relevant to the study of species evolution:

> Taken together, these observations suggest that the rapid evolution of COX II and cytochrome c prior to the divergence between humans and Old-World monkeys and the more recent rapid evolution of subunit IV that we present here are steps by which different parts of the complex modify or modulate parts of their function. (Hewett-Emmett et al. 1992, p. 5269, author's translation)

Complexity accompanies a molecular and cellular structural unity, built through time and in a world of possibilities. Biological evolution is in parallel based on this unity, but also on a molecular bricolage that leaves room for the unpredictable:

> In all likelihood, it is through the random joining of pre-existing DNA sequences that new genes may have been formed. (Jacob 1981, p. 73, author's translation)

Between physics, chemistry and biology, a set of functions, genetic rearrangements, as well as a flow of matter and energy, go through the body. The body, to maintain its self-preservation functions, consumes substances such as glucose, part of which is burned to release energy. Developments in organic chemistry are fundamental to understanding the formation and role of microstructures in larger organic ensembles. Moreover, the study of microorganisms could allow us to understand how the permanence of combinatorial biochemical mechanisms has traversed the history, since their origin, of certain structures of the living. Biochemistry would thus be a scale of stability in the biological complexity which permits for the reduction of cellular contents in physicochemical terms and the analysis of the reactions of the cell. Small changes redistribute the same structures in time and space and profoundly modify the forms and physiology[5].

By giving the rigor of quantitative methods to biology, genetics and biochemistry also give depth to organic unity: the scaffolding of living structures. After the rejection of a reductionist physics between the 18th and 19th centuries, biology became the science of an autonomous living being. In the middle of the 20th century, it gave back its place to physics in the context of the development of molecular biology, opening itself up even more to analysis and experimentation.

If "all this is translated by the words divergence, diversification, dispersion; one can no longer represent the succession of living forms through time by a table with a single column, or even with several parallel columns corresponding to independent series; the only figure that is suitable for describing the diversification of a group is the family tree" (Jacob 1976, p. 181, author's translation). This description goes beyond the framework of evolutionary theories to reach the heart of cellular and molecular phenomena. The multiplication of relations and possibilities is not only played out in a macroscopic spatiotemporal framework but also on microscopic scales, subject to variations in space and time. The living world, right down to the heart of its cells, is conceived in terms of phyla. Macroscopic and microscopic studies of life are correlated through the links between beings that lived in the past and those of today.

This spatiotemporal grid is objectified by Darwin, notably in its dual internal and external dimension:

> But in the great majority of cases, namely, with all organisms which habitually unite for each birth, or which often intercross, I believe that during the slow process of modification the individuals of the species

---

5 "[...] that it might require a long succession of ages to adapt an organism to some new and peculiar line of life, for instance, to fly through the air; and consequently that the transitional forms would often long remain confined to some one region; but that, when this adaptation had once been effected, and a few species had thus acquired a great advantage over other organisms, a comparatively short time would be necessary to produce many divergent forms, which would spread rapidly and widely, throughout the world" (Darwin 1896, p. 455).

will have been kept nearly uniform by intercrossing; so that many individuals will have gone on simultaneously changing, and the whole amount of modification will not have been due, at each stage, to descent from a single parent (Darwin 2021)

If the modifications appear to be gradualist, only varieties appear, leaving little room for biological novelty. Therefore, the transformation of one species into another would represent only the sum of small changes undergone by successive generations in the process of adaptation. The question of gradualism and saltationism is explored further by Goldschmitt. It proves to be heuristic for understanding the notion of a leap at the microscopic scale with effects visible at the macroscopic level.

Adaptation is the necessary result of this interplay between the biological system and its surroundings. This macroscopic perspective is reflected in the internal developments. From the scale of the internal mechanics of life to the study of the distribution of groups in space and time, an interpretation grid is woven that correlates the effects of the environment on the internal structure:

> The system only functions thanks to a series of successive transformations in which information intervenes. Entropy and information are as closely associated as the back and front of a medal. In a given system, entropy measures both disorder and our ignorance of the internal structure; information measures order and our knowledge. (Jacob 1976, p. 217, author's translation)

What does it mean to suppose an isomorphism of entropy and information? If the term "entropy" refers to unpredictability, "information" gives an account of a set of facts, of physical and mathematical determinations intended to understand nature. Entropy marks both the complexity in the potentiality of the living being and the complexity in its aspects relative to our understanding, whereas information gives an account of knowledge put into form according to physical and mathematical techniques. Reductionism, redeveloped in biology in the 20th century, was the basis for the concept of making legible biological information, as characterized by linear sequences of sub-units, bases in nucleic acids and amino acids in proteins. The genetic message, operating in terms of relationships between primary structures, codes and metabolic chains, thus became accessible. However, its development is neither linear nor totally predictable, insofar as the entropic dimension is nestled within relations with the environmen and possible influences of the maternal environment that are capable of generating accidents in the formation of tissues and organs.

Contemporary biology is marked by analogies between the living and the performing machine, insofar as the organization of the body, described in terms of cells and molecules, is understood according to networks for the exchange of information. However, life evolves and cannot be reduced to a machine. Indeed, biological disorder, in the sense of unpredictability, can result in certain errors, which can lead to mutations, which in turn can have consequences for the species by causing a change in the genetic message. However, once they have appeared, they are recopied and become invariant. If these errors emerge from the realm of contingency, they are inscribed by the evolution of invariance and replication. Although the genetic program is constituted by the combinatorics of essentially invariant elements, the rearrangements at the level of vital developments introduce zones of contingency. Complexity arises from the interactions between the stability of molecular components and development. Variation and mutation represent multiple possibilities of the genetic message. Heredity and generation are thus marked by a cascade of events that cannot be directed in a precise direction, neither by the environment nor by any component. Stability and variability are both carried by the genetic text, while chemical changes are blind and random:

> The genetic text is in perpetual reworking, it is unceasingly modified, corrected, adapted to reproduction in the most varied conditions, the successive retouching carried out by natural selection. Without thought to dictate it, without imagination to renew it, the genetic program is transformed by being realized. (Jacob 1976, p. 319, author's translation)

What about biotechnologies? Indeed, it is possible to question the links between theoretical biology and techniques through the perspective of an experimental reductionism at odds with the living. If biotechnologies model the applications of theoretical biology, do they not go in the direction of a reductionism limited to laboratory activities? What impact can they have on the living, conceived of as an exteriority to the experimental? Therefore, the more the techniques of experimentation progress, the more the living being seems to be able to be the object of modifications.

2.3. Reductionist biotechnologies?

According to Jacob (1976):

> [...] doing biology consists in studying the variations which occur in certain parameters in response to change, natural or provoked by another. Henceforth, all the efforts of biologists, their ingenuity, their approach, are aimed at finding a way to isolate one of the variables, to invent a technique for disturbing it in a calculated way, to measure the effects on the others." (p. 143, author's translation)

2.3.1. *From physics to biotechnology*

This description of 20th century biology, which aims to understand, modify and reproduce living things, can be seen as a result of the experimental physiology of the beginning of this same century. Let us take the example of the French physician and surgeon Alexis Carrel. In 1912, this scientist initiated a series of cell and tissue cultures, notably of cardiac tissue (Carrel 1912, pp. 516–530). These cultures proved to be suitable for continuation for several months at the rate of regular transplantations:

> Some are still living and have reached the beginning of the third month of their life in vitro. It was possible, therefore, to study the morphological and dynamic characters of tissues cultivated for more than two months outside of the organism. (Ibid., p. 519)

Next, in 1938, he changed the biological scale of his experiments, from the culture of cells to the culture of organs placed outside their original organism (Carrel and Lindbergh 1938). He participated in a technical reification of biological life transformed into an object of experiment, measurements and studies on the various physiologies, from the cell to the organ:

> The method of tissue culture is concerned with the cellular level of organization of living matter. The method of whole organ culture is concerned with the supra-cellular level. [...] The first method is related to the properties of tissues and blood cells as well as to the laws of cellular association during the manufacture of organs. The second method is related to the properties of organs and the laws of association of these organs within an organism. These methods are complementary. (Ibid., p. 5)

This deeply experimental project reflected a biomedical innovation that not only attempted to understand in order to reproduce cellular tissue and, more generally, functional mechanisms but also aimed to care for a fragmented body, notably through grafting.

Therefore, experimenting in a fractional way on sets of units gives a technological representation of a living being whose elements are measured and replicated within the framework of biotechnological developments. But apart from the biomedical and experimental fields of application, biotechnologies seem to have little influence on the forms of the living in their natural course. If techniques have entered into a societal reality, even into the ways in which we conceive of human nature, making the opposition between animal and machine aporetic, they do not seem to be able to contain the living in its potential variability and can therefore only give a fragmentary account of it. Indeed, to make the living a biotechnological object requires proceeding by scale and overcoming the difficulties linked to the interdependence of the functions in relation to each other; at the risk, in the contrary case, of damaging the biological objects of experience or of envisaging their applications in a limited way only.

With the development of physics and its ramifications during the 20th century in the fields of genetics and organic chemistry, new representations of a living being as a field for the development of new techniques have emerged. As this technicization of biological objects deepened the complexity of the relations between a living being in the laboratory and the living being as it is expressed in nature, it was necessary to think of the continuity between physics, chemistry and biology in order to understand the fundamental structures of biological phenomena. The physicist Erwin Schrödinger, in his book *What is Life?* (1967), highlighted the physical specificity of life at the level of its foundations while reintroducing a unity of physics and molecular biology. Therefore, the empirical regularity of ontogeny within phylogeny seems to be at odds with the probabilistic character of the microscopic physics of the elements that make up life. From physics to biology, we have seen that natural phenomena can be thought of in terms of the interactions between their different scales. This approach permits a certain continuity between the invariance of physicochemical elements and the variability of combinations. This reading grid makes it possible to consider shaping biological objects in their stable state. But a permanent questioning remains about the number and the nature of the physical elements, such as the atoms and the combinatorial activity of biology:

> How can we, from the point of view of statistical physics, reconcile the facts that the gene structure seems to involve only a comparatively small number of atoms (of the order of 1,000 and possibly much less), and that nevertheless it displays a most regular and lawful activity – with a durability or permanence that borders upon the miraculous? (Schrödinger 1967, p. 46)

If the life cycle of an organism shows an order that is at odds with that which can be observed in inanimate matter, this discontinuity is subject to scientific reduction through laboratory experiments. With his physicist's perspective, Schrödinger studied the fact that the links that unite the strong bonds between molecules in crystals are of the same nature as those that unite the atoms in a molecule. These studies permitted him to put unity back into discontinuity, this description of life going in the direction of a specific heredity made up of particulate atoms that are conserved even as the atoms separate and recombine.

Pauling (1960) also worked on the links between the concepts of quantum mechanics and molecular biology.

From the middle of the 20th century to the present, biotechnology, especially genetics, has focused on understanding how genes are expressed in organisms, determining their development in order to reproduce them. Therefore, studies on organ homology have changed scale to analyze the ways in which genes are expressed within parts of the organism. These explorations have highlighted the fact that gene homology does not determine the formation of similar structures and functions. Therefore, techniques to introduce a gene that determines the formation of the eye have been developed to verify how the same gene can be used differently or can, on the contrary, always produce the same functional structure:

> However, this approach has shown its limits, given that it can happen, for example, that the same developmental gene is recruited several times independently during the course of evolution, and controls the formation of structures without any homology. For example, a mouse gene (Pax-6), which plays an essential role in the development of the eye, has a homologous gene in Drosophila (eyeless) which also controls the formation of compound eyes. It has even been possible to introduce Pax-6 into the genome of a Drosophila, and to have it expressed at various sites in the body, thus causing the formation of compound eyes on the legs, antennae, etc. (Schmitt 2006, p. 431, author's translation)

These experiments on the links between molecular homology and organ homology have permitted us to understand how key genes express similar organs. By reproducing these phenomena, this experimentation was able to objectify the physicochemical mechanics, its impact in embryology and genetics and the fragility of the causal links that we infer in the developments of living organisms. The difficulties are in understanding the scope of the links between life and its foundations and how biology and techniques also seem to be bounded by the language they use. For example, the notion of genetic program refers to certainties, to mechanisms, all aspects of which are predictable. Although it is heuristic in the experimental framework, the term "program" seems to carry an intrinsic contradiction with natural biology; for example, at the level of the descriptions of the effects of the phenotype developments in relation to the genotype. The phenotype is defined at different scales (morphological, cellular and molecular), which are linked together. At the molecular level, it determines the cellular phenotype, which in turn determines the phenotype at the organism level. The proteins encoded by the genotype constitute the phenotype; however, several different genotypes can correspond to the same phenotype. Indeed, when a phenotype is determined by the expression of several genes, mutation of each of these genes results in the same phenotype. The causal demonstration that goes from genotype to phenotype is therefore tree-like, plastic and nonlinear. It is based on the analysis of the effects of changes in hereditary traits on individuals. The gene is a unit that introduces a discontinuity in the living world and determines the expression of phenotypic traits. If we extend the functional analogy between the mechanisms of heredity or development and the machine, it should appear that the elementary operations are always perfectly iterable. However, such combinatorial invariance seems to require the instruments and knowledge of biotechnologies to be obtained. If the intelligibility of the role of DNA goes beyond the notion of program, the latter seems necessary for the development of biological technologies.

The year 1972 was marked by the advent of genomics, a discipline that made it possible to study the various normal and pathological functions of an organism, or a cancerous process, at the level of all the genes of a species and no longer at the level of a single gene. The development of this discipline is linked to the development of technology in biology, but also to the development of computer science. Since the beginning of the 20th century, the complete sequencing of a growing number of genomes has been announced for various species, such as the *Caenorhabditis elegans* worm in 1998, the Drosophila fly or the dog in 2005. The sequencing of the human genome was announced between 2003 and 2006, and projects in synthetic biology to create the first synthetic genome are underway[6] (Dorrier 2016).

The physicalist, computer and technological frameworks question the complexity of life in attempts to understand it in a fractional way and renew the problem of its reducibility. The impacts of genomics, important for genetic medicine, contribute to giving a fixed and readable representation of the biology of the living. In genetics in particular, a framework of a priori logical deduction seems necessary to establish a causal reading that would allow us to understand how the genotype, which corresponds to all the characteristics of an individual, constitutes all the scales of the phenotype. The difficulties are linked to the real complexity of the physical systems studied by biology. It would be necessary to be able to isolate each mechanism, each action of a gene on a biochemical process, in order to understand its actions in a singular way. Biological experiments and technologies work on such fractionation.

---

6 Announced in 2005, the Human Connectome Project corresponds to the establishment of a map of the neural networks present in the animal and human brain. We then speak of "connectomic". These studies are being carried out on normal brains or brains affected by neurodegenerative diseases and follow the trajectory opened up by genomics.

2.3.2. *When the living extend beyond the experimental framework*

Goldschmidt's work on phenocopy[7], which was published in 1935, shows that, in teratogenesis, apart from technical intervention, modifications of the environment, in terms of temperature or chemical composition, lead to morphological and embryological modifications. Therefore, natural predictability is partially defeated by its interdependence on the environment. It is possible to deduce that the expression of the phenotype does not depend only on internal mechanisms, such as those of the genetic code or internal mutations, but also on the potential developments of the organism in interaction with its environment. In parallel with his studies on the effects of the environment on embryological processes, Goldschmidt defended the concept of saltatory evolution and elaborated on his theory of the "hopeful monster", thus stating the fundamental and unpredictable role of mutations in the context of adaptation to the environment:

> I used the term 'hopeful monster' to express the idea that mutants producing monstrosities may have played a considerable role in macroevolution. A monstrosity appearing in a single genetic step might permit the occupation of a new environmental niche and thus produce a new type in one step. [...] I think that this idea of the hopeful monster has come into its own recently. Only now is the exact basis of its evolutionary significance available. This basis is furnished by the existence of mutants producing monstrosities of the required type and by the knowledge of embryonic determination, which permits a small rate change in early embryonic processes to produce a great effect embodying considerable parts of the organism. (Goldschmidt 1940, pp. 390–391)

With Goldschmidt's observations, it becomes extremely complex to describe the action and function of a protein on a gene independently of its environment; internal constraints can no longer be considered as absolutely determinant for our biological system, nor can they be decoupled from external factors whose possibilities go beyond the "a priori conceivable". The slightest modification of the DNA, in the context of interdependence with the chemical or environmental context, can have important effects on the phenotype. These interactions of protein structures with the gene, and then with the environment, provided they are isolated, can be the subject of gene manipulation for therapeutic purposes.

---

7 The term "phenocopy" refers to a non-hereditary change in phenotype. It is described as being due to particular environmental conditions that simulate a phenotype similar to that resulting from a genetic mutation.

If biology as an expression of life goes beyond the notion of a program, which finds its application in laboratories and in projects for understanding, reading and caring for life, it influences, at the level of knowledge and increasingly precise exploitations that we make of it, a phenomenotechnique[8] that allows us to reproduce useful mechanisms for biomedicine. Indeed, the principles of programming imply the regularity of results. This regularity constitutes an apparent paradox with a living organism within which the greatest functional stability must give way to structural variability. However, it applies to the replication of biological phenomena in the laboratory context. Since the advent of molecular biology, every gene is considered as the carrier of information that permits the association of a biological function to it based on the chemical nature of the binding sites of the coded protein. However, there are ambiguities in the systematic association of a gene protein and a protein function. Plasticity and metabolic adaptability are concepts that limit this reduction:

> In the first analysis, then, this means that the gene is not an exact database and that the elementary operations that lead from the gene to the proteins are not perfectly iterable in the same way. (Miquel 2008, p. 197, author's translation)

Context plays a fundamental role in biology as soon as it is a question of describing a causal link that goes from the molecular level to the function. If the links are not linear, but flexible and elastic, in the framework of a natural biology, they are reified, calculated, measured in an experimental context:

> In the 1953 paper by Crick and Watson, replication seemed to be reducible to the property of nucleotides to pair up pair by pair for stereochemical reasons. We now know that this is no longer the case. (Miquel 2008, p. 230, author's translation)

The transition from epigenesis, a theory which since the 17th century has stated the idea that organisms are formed gradually from homogeneous and unorganized matter, to epigenetics has been fundamental for the transition from the study of organisms to the study of the behavior, at the microscopic level, of the elemental constituents. This change of scale has made it possible to determine the roots of the modifications in the mechanisms and functioning of the recombination or arrangement of the elements, and not only at the level of the genes themselves.

---

8 The Bachelardian concept of phenomenotechnics highlights the dependence of science on technique. When this term first appeared in 1931, it illustrated the implications of mathematical thinking.

This increasingly technical knowledge of the structures and organization of living organisms is the source of new therapeutic projects in which living organisms are fragmented and objectified. Indeed, therapies that intervene at the level of a "sick" gene are being developed and consist of importing a copy of a functional gene into a target cell, so that it is expressed there and leads to the production of the protein that is lacking. The gene added as a therapeutic element does not transform the diseased gene but is added to the genetic makeup of the cells to compensate for the deficient function. This is the treatment for single-gene diseases. These treatments are applied either directly in the organism or in the laboratory to the targeted cells before they are injected into the patient. The second possibility allows one to work in a fractionated way, with better control of the steps, and to avoid the dispersion of the treatment in non-targeted organs. This solution is most often used for the treatment of blood diseases, given the simplicity of the medical procedure, which consists of taking the cells to be corrected through a blood sample. Gene therapies experienced a new boom around 2007 in the context of the treatment of the *xeroderma pigmentosum* pathology, the sufferers of which are known as "moon children". The biotechnical project is to introduce into the cells of a patient the normal "version" of the gene whose alteration is responsible for the disease. These treatments then target the stem cells, which are responsible for the production of skin cells. In the historical extension of the experimental program opened by Carrel on cell and organ cultures, the treatment consists of isolating and cultivating stem cells, re-generating skin in the laboratory and performing transplants of this skin that would have been "grown" in the laboratory. These therapies are part of the context of controlled living within the limits of the laboratory.

The recombinant DNA technique developed by Paul Berg in 1972 (Berg and Singer 1976), which involves a deoxyribonucleic acid molecule created in the laboratory and composed of nucleotide sequences from several sources, permits the creation of sequences that do not exist in living organisms. Recombinant technology is widely used for the production of therapeutic proteins (such as insulin).

As a set of techniques that allow for the genetic constitution of an organism to be modified, genetic engineering permits, for example, the transfer of genetic material, also called transgenesis, from one organism to another. First performed by Herbert Boyer and Stanley Cohen in 1973, this technique made it possible to produce artificial insulin in 1978. The field of genetic engineering carries representations of the control of the genome and the possibility of intervening technically, although there are uncertainties about its effects on the complexity of the human genome and the interdependence between all genes. CRISPR-Cas9[9] technology is a medical innovation developed in 2012 by the French scientist Charpentier (2015,

---

9 Cas9 is a protein of bacterial origin with antiviral properties. Its ability to cut DNA at specific sequences has made it a molecular biology tool with broad prospects for use.

pp. 363–365) and her American colleague, Jennifer Doudna, that uses molecular scissors to cut out certain genes.

Therefore, a range of biotechnologies are being developed to eliminate or repair altered genes directly in the cell. In this context, genome editing allows for the targeted repair of genetic mutations by using several elements directly at the cellular level, such as nucleases to cut the genome where necessary, or a segment of DNA that is used to repair the genome and restore a functional gene. These techniques progressively modify the living being in its natural course, thus entering into interactions with the notion of replication or invariance. In what way does the modified object integrate into the living? One can question its possible redeployment in nature.

The new technologies that have the living world as their field of exploration also have anthropological stakes, at the level of the representations that humans have of life, as well as in the perspective of their inclusion in our societies. We have seen, previously, that it is difficult to reconcile the reductionism associated with the concepts of programs and machines with the complexity of living beings described in terms of self-organization, emergence or inventive adaptation.

Reductionism and complexity seem to shape two distinct but articulated and complementary fields of research around the notion of life:

> Reductionism is opposed to the idea of complexity and *vice versa*. This antagonism explains the difference in vision of the research communities, and it justifies the idea of incommensurability of practices. However, this antagonism already simplifies the reality: first, because there is reductionism only with respect to complexity, in other words, the two disciplines develop different strategies to solve the same problem, that of the complexity of living organisms [...]. (Ujéda 2016, p. 77, author's translation)

If complexity and reductionism represent two sides of the same coin, they are able to articulate themselves from the interactions between biotechnologies and theoretical biology. Biology in its technical dimensions reduces the complexity of the living to the description of its elements in order to understand the mechanics of the living and to imitate it in its therapeutic perspectives, while theoretical biology offers a range of variable combinatorial movements on which to experiment. If life, in its spontaneous manifestations, overflows the experimental framework by its complexity, it becomes the object of biotechnologies in perspectives of care and explorations.

The complexity described when we approach the living is taken in a network of historical stages that have participated in the construction of biological knowledge and the expression of its phenomena, as much at the level of the invariance of the physical and chemical components as in the variations of their expression. Biology is thus marked by complexity in its activity of updating its possibilities, in its morphological and functional variability and in its relationship to environmental modifications. Finally, biotechnologies renew a questioning of the inscription of new forms of life, imitated and artificial, in the natural mechanics of life.

2.4. References

Aldini, G. (1804). *Essai théorique et expérimental sur le galvanisme, avec une série d'expériences faites en présence des commissaires de l'Institut national de France, et en divers amphithéâtres de Londres*. Fournier fils, Paris.

Andrault, R. (2012). Notes de lecture : François Duchesneau, Leibniz, le vivant et l'organisme (Paris, Vrin, 2010). *Philosophiques*, 39(1), 295–305.

Angleraux, C. (2018). La simplicité monadique chez Leibniz. *Philonsorbonne* [Online]. Available at: https://journals.openedition.org/philonsorbonne/945 [Accessed 8 January 2022].

Aristotle (2009). *On the Soul*, translated by J.A. Smith [Online]. Available at: http://classics.mit.edu/Aristotle/soul.html.

Bacon, F. ([1637] 1945). *Histoire de la vie et de la mort*. La Boëtie, Brussels.

Berg, P. and Singer, M. (1976). Recombinant DNA: NIH guidelines. *Science*, 193(4249), 186–188.

Bernard, C. ([1885] 2022). *Leçons sur les phénomènes de la vie, communs aux animaux et aux végétaux*, 2nd edition. Hachette, Paris.

Bichat, M.F.X. (1799–1800). *Recherches physiologiques sur la vie et la mort*. Brosson, Gabon et Compagnie, Paris.

Buffon, G.-L.L. (1888). *Histoire naturelle des animaux*. H. Lécène, H. Oudin, Paris.

Carrel, A. (1912). On the permanent life of the tissues outside of the organism. *The Journal of Experimental Medicine*, XV, 516–530.

Carrel, A. and Lindbergh, C. (1938). *The Culture of Organs*. Hamish Hamilton, London.

Chapouthier, G. (2018). Les racines de la complexité en mosaïque. *Revue philosophique de la France et de l'étranger*, 143, 3–10.

Charpentier, E. (2015). CRISPR-Cas9: How research on a bacterial RNA-guided mechanism opened new perspectives in biotechnology and biomedicine. *EMBO Mol. Med.*, 7, 363–365.

Cherici, C. (2016). *Anatomophysiologie du cerveau et du cervelet chez Vincenzo Malacarne (1744–1816). Édition critique des textes de l'Encefalotomia nuova universale (1780) ; Vera esposizione della struttura del cervelletto umano (1776).* Hermann, Paris.

Cherici, C. (2020). *From Clouds to the Brain.* ISTE Ltd, London, and Wiley, New York.

Crick, F. and Watson, J. (1953). Molecular structure of nucleic acids: A structure for deoxyribose nucleic acid. *Nature*, 171, 737–738.

Darwin, C. ([1875] 2008). *The Variation of Animals and Plants Under Domestication.* John Murray, London [Online]. Available at: https://www.gutenberg.org/cache/epub/24923/pg24923-images.html.

Darwin, E. ([1796] 2020). *Zoonomia; or the Laws of Organic Life.* J. Johnson in St. Paul's Church-Yard, London [Online]. Available at: https://www.gutenberg.org/cache/epub/15707/pg15707-images.html.

Darwin, C. ([1859] 2021). *On the Origin of Species by Means of Natural Selection.* John Murray, London [Online]. Available at: https://www.gutenberg.org/files/1228/1228-h/1228-h.htm.

De Vries, H. (1909). *Espèces et variétés – Leur naissance par mutation.* Alcan, Paris.

Deutsch, J. and Le Guyader, H. (1995). Le fond de l'œil : l'œil de la drosophile est-il homologue de celui de la souris ? *Médecine/Sciences*, XI, 1447–1452.

Dorrier, J. (2016). Writing the first human genome by 2026 is synthetic biology's grand challenge [Online]. Available at: https://singularityhub.com/2016/10/10/writing-the-first-human-genome-by-2026-is-synthetic-biologys-grand-challenge/ [Accessed 12 February 2022].

Duchesneau, F. (1982). *La physiologie des Lumières. Empirisme, modèles et théories.* M. Nijhoff, La Haye.

Galilei, G. ([1623] 1968). *Il Saggiatore, nel quale con bilancia squisita e giusta si ponderano le cose contenute nella Libra.* Opere di Galileo Galilei, Florence.

Galvani, L. (1953). *Commentary on the Effect of Electricity on Muscular Motion*, translated by Green, R.M. Elizabeth Licht, Cambridge.

Goldschmidt, R.B. (1940). *The Material Basis of Evolution.* Yale University Press, New Haven.

Grossman, L.I., Hewett-Emmett, D., Lomax, M.I., Yang, T.L. (1992). Rapid evolution of the human gene for cytochrome c oxidase subunit IV. *Proc. Natl. Acad. Sci.*, 89, 5266–5270.

Hooke, R. (1665). *Micrographia: Or Some Physiological Descriptions of Minute Bodies Made by Magnifying Glassees: With Observations and Inquiries Thereupon.* Jo. Martyn and Ja. Allestry, London.

Jacob, F. (1976). *La Logique du vivant ; une histoire de l'hérédité.* Poche, Paris.

Jacob, F. (1981). *Le Jeu des possibles : essai sur la diversité du vivant.* Fayard, Paris.

Jussieu, A.-L. (1778). Exposition d'un nouvel ordre des plantes adopté dans les démonstrations du Jardin Royal. In *Mémoires de mathématique et physique de l'Académie royale des sciences de Paris, 1774*, Paris, Imprimerie Royale.

Kant, I. ([1790] 1965). *Critique de la faculté de juger*, translated by Philonenko, A. Vrin, Paris.

Kauffman, S.A. (2019). *Au-delà de la physique, l'émergence de la vie*. Dunod, Paris.

Lamarck, J.-B. (1815–1822). *Histoire naturelle des animaux sans vertèbres*. Verdière, Paris.

Lamarck, J.-B. (1830). *Système analytique des connaissances positives de l'homme*. Baillière, Paris.

Leibniz, G.W. ([1714] 1846). *Monadologie, Œuvres de Leibniz*. Charpentier, Paris.

Linnaeus, C. (1735). *Systema naturae per regna tria naturae : secundum classes, ordines, genera, species cum characteribus, differentiis, sinonimis, locis*, vol. I. Holmiæ, Stockholm.

Linus, P. (1960). *The Nature of the Chemical Bond*, 3rd edition. Cornell University Press, Ithaca.

Malacarrne, V. (1803). *Della esistenza di molti sistemi e della influenza loro nella economia animale*. Stampa nel Seminario, Padua.

Malacarrne, V. (1811). *Di mostri umani, lezione accademica terza del professore Vincenzo Malacarne da Saluzzo : conferma della proposizione circa alla produzione de'mostri umani*, vol. 15. Società italiana delle scienze, Rome.

Mendel, J.G. ([1865] 1907). Recherche sur les hybrides végétaux. *Le Bulletin Scientifique de la France et de la Belgique*, 41, 371–419.

Miquel, P.-A. (2008). *Biologie du XXIe siècle : évolution des concepts fondateurs*. De Boeck Université, Brussels.

Monod, J. ([1970] 2014). *Le hasard et la nécessité. Essai sur la philosophie naturelle de la biologie moderne*. Points, Paris.

Morange, M. (1994). *Histoire de la biologie moléculaire*. La Découverte, Paris.

Pittendrigh, C.S. (1958). Adaptation, natural selection, and behavior. In *Behavior and Evolution*, Roe, A. and Gaylord Simpson, G.G. (eds). Yale University Press, New Haven.

Rostant, J. (1958). *Aux sources de la biologie*. Gallimard, Paris.

Sanger, F. (1959). Chimie de l'insuline : la détermination de la structure de l'insuline ouvre la voie à une meilleure compréhension des processus vitaux. *Science*, 129(3359), 1340–1344.

Schmitt, S. (2000). L'œuvre de Richard Goldschmidt : une tentative de synthèse de la génétique, de la biologie du développement et de la théorie de l'évolution autour du concept d'homéose. *Revue d'histoire des sciences*, 53(3/4), 381–400.

Schmitt, S. (2006). *Aux origines de la biologie moderne – L'anatomie comparée d'Aristote à la théorie de l'évolution*. Belin, Paris.

Schrödinger, E. (1967). *What is Life? The Physical Aspect of the Living Cell*. Cambridge University Press, Cambridge.

Serres, E.R.A. (1824). *Anatomie comparée du cerveau dans les quatre classes d'animaux vertébrés appliquée à la physiologie et à la pathologie du système nerveux*. Gabon et Compagnie, Paris.

Serres, E.R.A. (1859). *Anatomie comparée transcendante. Principes d'embryogénie, de zoogénie et de tératogénie*. Didot frères, Paris.

Ujéda, L. (2016). Étude philosophique de la biologie de synthèse : pour une analyse de la complexité des biotechnologies en société. PhD thesis, Paris-Est University, Paris.

Wolff, C.F. (1759). *Theoria Generationis*. Hendel, Halle.

3

Two Complexities: Information and Structure Content

3.1. The simple, the random and the structured: A triangle of concepts key to a complete understanding

When we consider the different stages in the evolution of the universe, as described by modern cosmology, we are tempted to say that it is becoming "more complex" (since the Big Bang at least). In the same way, it is said that the Earth's ecosystems include more and more numerous and varied elements, interacting in increasingly subtle and delicate ways. On a smaller timescale, we still talk about the "complexification" of the human societies, social and commercial exchanges, technological objects and scientific theories. But what exactly is complexity? What precisely is increasing when we speak of increasing complexity? Can we measure or at least define this "complexity" in a rigorous manner to define it as a mathematical and scientific concept?

The issue has long preoccupied researchers, and a multitude of methods and ideas have been proposed for defining and assessing the complexity and organization of objects, living things, processes and interactions (Mitchell 2009; Grassberger 2012). Although as of yet no unanimous agreement has emerged, one of the proposals – Charles Bennett's logical depth – has received widespread attention (Bennett 1988, 2012; Gell-Mann 1995; Delahaye 1999, 2010; Li and Vitányi 2008; Mayfield 2013). With results accumulated year after year since its introduction in 1977, it is considered the most serious avenue for a robust scientific definition of the idea of "structural complexity" or "richness in organization". The development of tools that make it possible, at least in some cases, to measure it leads to the consideration of other applications.

Chapter written by Jean-Paul Delahaye.

3.2. Calculation, the key to the solution

Not surprisingly, the theory of computation is at the heart of the definition proposed by Bennett (1988). What is more surprising is that the proposal to define structural complexity with the help of this mathematical theory was formulated by a physicist otherwise known for his research in quantum mechanics and thermodynamics[1]. If this "computational" formulation had been proposed by a computer scientist or a mathematician, it might not have received so much attention: the fact that a physicist felt the need to resort to computational theory suggests that Bennett's formulation, based on the notion of a program, is based on profound arguments and is not just a way for computer scientists to show off!

To express the idea in a sentence that will be refined later, Bennett proposes to consider that the structural complexity of an object Ob (which he calls *logical depth*) is the computational effort measured in machine cycles (or more generally in computational resources: time, memory or parallelism) such that, starting from a short description of the object Ob – a program – a perfectly detailed explicit description of Ob is produced.

Sometimes the complete description of an object is produced quickly by a short program. This is the case, for example, for a block of crystal or for a periodic numerical sequence like 010101...01 (one million times 01). It is then clear that it is not a complex object. Objects with a simple structure have short minimal programs, and these produce their results without requiring large computational resources. Such objects will have a low logical Bennett depth. However, there is a second class of poorly structured objects. A sequence s indicating the result of a million coin flips is weakly organized and therefore not structurally complex. A short program producing the writing of such a sequence will necessarily be in the form "print s" because one cannot compress, except in an insensitive and exceptional way, what is random. A "print s" program runs quickly, since it contains no loops or long parts to run: to run it is to traverse it. A purely random object will therefore – like the sequence 010101...01 (a million times 01) – be quickly produced by the shortest programs that give a complete description of it. It will therefore have a low Bennett logical depth.

To be authentically organized or structurally complex, an object must be able to be described quite briefly by a program or something similar to it, which takes a fairly long time to run. There are good reasons to believe that this is the case for a microprocessor, a city or a mammal, which are among the most structurally complex objects we know.

---

1 See: https://scholar.google.com/citations?user=mkjGmJEAAAAJ&hl=en.

When we attribute a great structural complexity to an object, it is because we see, or believe we see within in it, the traces of a long causal history comparable to a calculation. This history has forged the relations that its different parts maintain with each other and that constitute its organizational complexity. If one accepts the idea that it is far more likely that the object one qualifies as structurally complex comes from a long but relatively simple to define dynamic, rather than from a brief process that would have generated it by chance, then Bennett's idea of taking into account the computation time of the shortest programs that generate the object appears natural and in accordance with common sense.

3.3. Thought experiment

Before specifying how Bennett's idea translates mathematically, let us formulate one more argument in its favor with a thought experiment. Imagine that we discover written on a tree trunk a billion binary digits of the number π materialized by long (for the 1s) and short (for the 0s) lines. What are we supposed to think about this?

– This sequence, which is only one among $2^{1\,000\,000\,000}$ others of the same length, is there completely by chance, due to bad weather or the blind work of some insects. It is pure chance that these strokes accurately reflect the first billion binary digits of π.

– A particular process – physical or involving humans and perhaps a computer – was responsible for the calculation of the billion digits copied there. In this case, the origin of the digits would be attached to a rather short program, or what amounts to the same thing, to some physical laws, the description of which would be brief in comparison to the billion symbols written on the tree trunk.

If you opt for the first choice, you choose a conception close to that of the biologists who, before Louis Pasteur, defended the idea of *spontaneous generation*, maintaining, for example, that rats are sometimes born from a pile of abandoned wet rags without any rat having previously passed through it. The second hypothesis corresponds to the more reasonable idea – at least in the eyes of modern science – that the origin of rats in a pile of rags is the same as that of other rats: they were born of parents who passed through there, and in the end, like them, they are the product of biological evolution on Earth which, for more than 3 billion years, has elaborated living forms with increasingly rich and delicate structures. This evolution works by using the laws of physics and chemistry that make atoms and molecules interact, all producing a form of calculation that is both parallel and extremely long.

For the rats in the rag pile, as with the billion digits of π in our thought experiment, the structural complexity we discover obviously comes from a long, but quite briefly definable, causal process. In computer terms, we would say that these objects are the result of the computation of "a short program that has run for a long time".

Provided that we adopt a broad conception of what a computation is, Charles Bennett's idea corresponds to the second option, and it seems to be the right one. This broad conception of computation is quite easy now, since the developments of computer science show us on a daily basis that any dynamic process can be simulated by a computation or, conversely, can be a means of performing computations.

3.4. Mathematical definition

Charles Bennett's general idea is made mathematically precise in several ways, the most elementary of which easily lead to theorems. The first mathematical version of Bennett's definition consists of positing that the logical depth $P(s)$ of the finite sequence s of 0 and 1 (any physical object of finite size can be reduced by digitization to such a sequence) is the computational time (measured by the number of elementary computational steps) that the shortest program (noted as s^*) that produces s executes to arrive at s (if there are several programs that are equal in size, the fastest one is selected). Of course, the programming language used to write the programs we are talking about is important and changing it will change the definition. However, as long as one chooses a fairly general language (those used in computer science are C, Java, Basic, Python, etc.), Bennett shows that his proposed definition of structural complexity only varies a little. This "invariance theorem" is obviously central, and it is thanks to this that the notion of logical depth can be taken seriously.

In practice, determining the shortest program s^* to generate s is very difficult and cannot be done systematically. It has been shown that the passage from s to s^* is indeed non-computable: there is no algorithm allowing for any s to know without error s^* and its computation time $P(s)$. This algorithmic undecidability comes from that demonstrated in 1936 by Alan Turing, which concerns the halting problem in programs: no method in analysis makes it possible to know in advance if a program will start to loop.

However, one should not draw from this impossibility of a systematic exact calculation for any s of $P(s)$ the conclusion that one cannot evaluate $P(s)$. Data compression algorithms are tools that compute approximate versions of s^*, and their use opens up the possibility of evaluating $P(s)$.

Similarly, the Kolmogorov complexity of the sequence s, which is the length (measured in bits) of s^*, is not computable; however, because of data compression algorithms, it can be evaluated. This has led to various applications in the field of the automatic classification of texts, musical scores and genetic sequences (Delahaye 2010). The non-computability that has sometimes been wrongly seen as an obstacle to the practical use of Kolmogorov complexity and Bennett's logical depth is in fact surmountable because of data compression algorithms.

3.5. Random complexity and structural complexity

Let us specify that the Kolmogorov complexity of a sequence, which is a very important notion in theoretical computer science (and now in physics, psychology and biology), could not be this measure of "structural complexity" that gives the sense of rigorousness to the idea of complexification evoked in cosmology and biology. Indeed, a sequence s that has been flipped has a maximal Kolmogorov complexity (among the sequences of the same length), whereas, as we have already noticed, no structure is present in general.

Kolmogorov's complexity measures the "information content" of a sequence s, that is, the minimal quantity of information that must be memorized in order to be able to reproduce s, even if this information is of no interest, as is the case for tossing a coin. The structural complexity – which is evoked when we speak of complexification and which Bennett's logical depth evaluates – is linked to the richness of organization, to the number of correlations and interactions, apparent or hidden, between the parts of s, and to the quantity of interdependencies between the parts, all the things that are absent in a random sequence of "0" and "1".

This distinction that we all make without necessarily being aware of it between random complexity (or incompressible information content that is measured by the Kolmogorov complexity, the size of $s*$) and structural complexity (that which Bennett wants to evaluate, and which, to a first approximation, is the computation time of $s*$) is essential for the progress of a general theory of complexity. It is because this was not understood earlier, and because it was not realized that the impossibility of an exact calculation of the formal versions of these notions (Kolmogorov's complexity and Bennett's logical depth) did not prohibit their use, that some sought, and that others still tend to seek, to formulate particular definitions of complexity that only apply within limited contexts. A general – sometimes called universal – theory of complexity that makes a clear distinction between "random complexity" and "structural complexity" is possible and practicable, and, as Bennett has proposed, finds its basis in the theory of computation.

More advanced and more precise mathematical versions of Bennett's notion of logical depth have been proposed. These versions do not just involve the shortest program producing s but also take into account all programs producing s that cannot be compressed by more than k bits (we will note this notion $P_k(s)$). Sometimes we even consider all programs producing s but assign a greater weight to the shortest programs, which are considered more likely origins of the object s than the long programs. These versions have better properties than $P(s)$, including permitting the demonstration of a "slow growth law", which indicates that the structural complexity cannot grow rapidly. Unfortunately, these definitions, which are more satisfying on an abstract level, are more difficult to apply than the basic definition that we have noted $P(s)$. For details on these points, see Bennett's original paper (1988), which remains the best technical paper on the subject, followed by the review by Li and Vitányi (2008).

3.6. Recent progress

Although progress in the understanding and use of Bennett's logical depth has been rather slow since his definition, it should be seen as a confirmation that Bennett's idea is satisfactory and that, even if nothing is very easy in this field subject to many "undecidabilities", it is necessary to continue his study, which, in any case, if only in an abstract way, helps us to grasp the nature of the complexity for which our intuitive understanding must now become mathematical and scientific.

Among the attempts to define a measure of structural complexity, some are derived from the theory of computation while being different from Bennett's concept. These other notions are in competition with Bennett's idea and it is therefore necessary to understand whether the notions are related.

The notion of "sophistication" introduced by Koppel and Atlan (1987) and the notion of "effective complexity" developed by Seth Lloyd and Murray Gell-Mann (Gell-Mann 1995; Li and Vitányi 2008; Mitchell 2009) are based on the idea of a possible separation between the random and organized components of an object. The idea behind the definitions proposed by these researchers in mathematical forms that sometimes obscure their naturalness is as follows: when one examines a program producing an object, it is possible – at least in some cases – to clearly recognize two parts; one part represents what fixes and describes the structure of the object; the other is what dresses up the structure with elements of lesser importance that are sometimes even random.

Consider for example a finite sequence s of pairs of binary digits 00 or 11 whose succession is random:

$s = 00\ 11\ 11\ 00\ 00\ 00\ 11\ 00\ 11\ 00\ 00\ 11\ 00\ 11\ 11$

The shortest $s*$ program producing this sequence will have two parts that are fairly easy to identify, as they will be in the form:

"print the sequence 011000101001011 by doubling each number".

The regular structure of s is expressed by the "print by doubling each digit" part. The disordered component is found in the descriptive and incompressible part of 0 and 1.

When dealing with a more structured sequence s than the one in our example, the length of the non-random part will be greater. It is this length which, according to Koppel, Atlan, Lloyd and Gell-Mann, measures the organizational richness – the structural complexity – of the sequence s.

The part of the program where the sequence has been written without doubling the digits 011000101001011 is all the longer given that the sequence s contains many random elements. This is the random part of the sequence, and its length is roughly the Kolmogorov complexity of s.

Technical difficulties make the mathematical formulation of this idea, of splitting a short program into two parts that result in s, tricky; nevertheless, what we have just explained is the idea of *sophistication* and *effective complexity*. Since these definitions are arguably as natural as Bennett's definition of logical depth, the question has been posed of whether there is a connection between these two seemingly quite different views of structural complexity. An answer was given in 2010 in a remarkable paper by Ay et al. (2010) of the Max Plank Institute in Leipzig. The researchers formally establish that high effective complexity always leads to high logical depth, thus confirming the validity of Bennett's definition.

3.7. Less undecidability

Recent research has led to a second advance in the theory which comes from the development of computable versions of logical depth. These versions make simplifying and restrictive assumptions about computational processes but result in definitions that are free from the troublesome non-computability of $P(s)$ (see Antunes et al. 2009; Antunes and Souto 2010).

A major argument for those who think that Bennett's concept of logical depth, or related concepts, is key to understanding the real world has been to see it recognized in scientific work related to that which in the world is a kind of grand calculation: biological evolution. Danchin (1998) from the Institut Pasteur, Baptist (2012) from the University of Grenoble, Collier (2008) from the University of Durban in South

Africa and Mayfield (2013) and Gaucherel (2014) all independently defend the idea that Bennett's logical depth is a useful and important concept in biology. For them, it makes it possible to avoid certain overly naive considerations about the complexification of living beings and it sheds light on the idea of progress in the lineages of living organisms – progress that would only be the accumulation of computational results (see also Delahaye and Vidal 2018).

3.8. Experimentation

Among the latest advances in the demonstration that Bennett's logical depth is the right concept to define and measure structural complexity, various works of an experimental nature have led to the beginnings of a practical validation of Bennett's idea, the intuitive foundations of which are not in doubt but for which the potential for application has yet to be shown.

The first work consisted of measuring the computation time of the shortest program producing an image. As performed by Zenil et al. (2012), the method evaluated the decompression time required for images representing various more or less structured objects. This decompression time can be equated to $P(s)$, since the compressed file of s is a version of s^* and it follows that the computation to go from the compressed file to the complete file is comparable to the computation that s^* performs to produce s, that is, $P(s)$. The results of these computation time measurements were used to order the images by increasing structural complexity. The resulting classification was close to the expected one, and it placed as expected at the beginning of the ranking the perfectly simple or totally random images, and at the end of the ranking, the images of the most structurally rich objects (see section 3.9.5).

The second work, based on the massive enumeration of small Turing machines, was performed by Soler-Toscano et al. (2013). It permits us to evaluate, for short sequences of 0s and 1s, the Kolmogorov complexity and Bennett's logical depth. These calculations show, as expected, that the two notions of complexity are independent and that the Bennett logical depth is small for both simple sequences (10×0, for example) and random-looking sequences. Again, the coincidence between theory and approximate calculus shows that, despite the theoretical undecidability of the notions, one can try to do something with them, and that what one finds then is in accordance with what the theory suggests.

The idea of complexification, the topic of many commentaries and just as many speculations, is becoming a scientific and mathematical subject based on a rigorous general theory and which, in the simplest cases, produces measurement tools. One can even dream of reaching a deep theoretical understanding in terms of the development of the complexity of the universe and living beings, which, as with what happened with the idea of spontaneous generation, would render definitively obsolete the elucidations of creationists and the notion of intelligent design.

Beyond all this, we, together with Clément Vidal, have proposed a complexity ethic that builds on Bennett's concept of logical depth (Delahaye and Vidal 2018, 2019). It attempts to define a conception of moral value independent of any anthropocentric view. This ethic unifies and generalizes many values that are now universally adopted: the protection of nature, the safeguarding of endangered species and the respect for human life and all forms of richness in structures, and thus, in particular, for all living beings and all artistic or scientific intellectual productions.

3.9. Appendices

3.9.1. *Complexification*

The universe on a large scale seems to be becoming more and more complex, and more and more varied forms of organization are emerging and interacting with each other. Life on Earth is one of these forms.

Complexification is not necessarily monotonic and can sometimes temporarily decrease: there can be sudden destruction of complex structures which disappear definitively. However, at a larger scale, we observe in the universe a complexification of the structures present, in particular, of course in the living world. In the world of technology (cars, telephones, computers, etc.), the evolution is also obviously a progressive complexification, even if sometimes the better designs of some devices over their predecessors are simplifications. The conception of a universe becoming more complex is not a finalistic conception but is only the one observable at a large scale for which we seek to understand deeper meanings. For further reading on this subject, see Delahaye and Vidal (2018).

The question of whether we can provide a meaning other than an intuitive one to this observed trend toward more *structural complexity* requires that we first satisfactorily define this notion and that we do not confuse it with *random complexity* (the information content), which is the worthless complexity produced, for example, by the toss of a coin.

Charles Bennett has proposed a definition of this structural complexity (which would be at best slowly increasing; see section 3.9.4) based on the theory of calculation. The idea is that structurally complex things are those whose origin can only be traced to a long process of maturation and evolution, a process that can be likened to a series of calculations. Bennett asserts that structurally complex things are those whose definition (by a program) is relatively brief but whose production (by the same program) requires a long time to calculate. *Bennett's logical depth*, which proposes to measure the structural complexity of a digitized object s, is, in its simplest version, the computation time of the shortest program s^* that produces s.

3.9.2. *Random and structural complexity*

Schematically, there are four kinds of digital objects:

– *Simple objects without complicated structures*: a sequence of "01" a million times; the tiling of a plane with a periodic pattern; a crystal lattice. These objects, once digitized, are described by short programs whose operation is rapid.

– *Random objects without any particular organization*: a gas, a series of results from the lottery draw or heads or tails. It takes a lot of memory to store their descriptions because, having no structure, one can do no better than enumerate their elements without having a description shortcut available. These are incompressible and no subtle calculation is needed to produce them from their shortest description, which is nothing but themselves, or something close to it.

– *Richly structured objects with no random component*: the sequence obtained by a recurrence relation like $x(0) = 1$, $x(1) = 1$, $x(n+2) = x(n)+x(n+1)$, the drawing of a memory chip, the plan of a building like the Eiffel Tower. Nothing is random, and a clever enough observer, by identifying the structure, will be able to deduce a fairly brief definition.

– *Richly structured objects with random elements*: a sequence with 10,000 repetitions of a heads or tails sequence 40 digits long: 0111001011100101010 100111101010001001010.

Of course, in the real world, most objects belong to the fourth category. A living being is very strongly structured but at the same time has many random elements (the emplacement of hairs on a dog or the detail of a leopard's spots are largely random). In a real piece of furniture, there is both structure (fixed by the plane of the piece of furniture) and randomness; for example, in the design of the wood's fibers.

Kolmogorov complexity (or random complexity) measures the amount of randomness present in an object. It is its incompressible information content, that is, the size of the shortest program capable of producing a complete (lossless) description of the object.

Bennett's logical depth (or structural complexity) is intended as a measure of the level of structure present in the object. According to Charles Bennett, it is measured by the computing time of the shortest program producing the object or more generally by the computing resources necessary for a short program to produce the object.

3.9.3. *Incalculable but approximate*

The mathematical notions of Kolmogorov complexity $K(s)$ and Bennett's logical depth $P(s)$ are both non-computable: there can be no algorithm for object s provided in numerical format to calculate $K(s)$ and $P(s)$.

This non-computability has been misunderstood and it has often been concluded that these two notions of complexity drawn from the theory of computation are impossible to apply in practice. This is false, and applications of Kolmogorov complexity have been proposed in biology (to calculate phylogenetic trees), linguistics (to classify language families), comparative literature (to classify texts), music (to group pieces by the same composer), etc. (see Li and Vitányi 2008; Delahaye 2010).

These applications use data compression algorithms with the idea that if a compression algorithm identifies the main regularities of a (numbered) object, then the size of the compressed file will be a satisfactory approximate value of the Kolmogorov complexity of the object.

In computer science, we constantly encounter non-calculable problems that we solve in an approximate way. The fact that $K(s)$ and $P(s)$ are not computable does not prevent us from approaching them in the following sense: there is a sequence of algorithms $Algo_1$, $Algo_2$, etc., such that for any sequence s, the results $Algo_1(s)$, $Algo_2(s)$, etc., converge to $K(s)$. Likewise for $P(s)$.

When running these algorithms and stopping on one of them, for example the thousandth, we do not know if we are very close to $K(s)$ (or to $P(s)$), but we know that we are on the "right path". Such methods are in practice capable of producing interesting and useful results. Their strength comes from the fact that the data compression algorithms (which have been carefully perfected for more than 50 years) seek out regularities within the data submitted to them, which is exactly the work needed to calculate $K(s)$ and $P(s)$. For more on these subjects, see Li and Vitány (2008) and Delahaye (2010).

3.9.4. *The law of slow growth*

The idea that objects with a rich organization cannot suddenly arise from nothing is natural. Charles Bennett, regarding this idea, speaks of the "law of slow growth". A good measure of structural complexity must have the property of not increasing suddenly during a dynamic process, because the structure content cannot increase suddenly without a lot of calculations having been performed. Logical depth is a measure of both structural content and computational content. Bennett (1998) was able to demonstrate that logical depth verifies the law of slow growth. This is a strong argument in favor of the idea that Bennett's definition of logical depth does indeed measure structural complexity.

Eukaryotes, multicellular organisms, vertebrates, mammals and humans appeared gradually and slowly on Earth. Similarly, the history of calculating machines has been made at a limited speed: the path from Pascal's machine to smartphones is inconceivable without numerous and progressive intermediate stages. While Kolmogorov complexity, which measures information content, can increase suddenly (a crystal thrown on the ground, a sheet of white paper that one crumples), what is richly structured never arises quickly: there is no spontaneous and sudden generation of organized complexity.

3.9.5. *Experimental evaluation of K(s) and P(s)*

A data compression program can evaluate both Kolmogorov complexity and Bennett logical depth. The idea is that the compressed version of a file should be seen as a short program that spawns the file. Its size therefore indicates an approximate value for the Kolmogorov complexity. In addition, the time necessary for the decompression of the file is comparable to the calculation time required by this short program to produce the initial file and thus to an evaluation of its logical depth.

We obtained practical confirmation of this idea by classifying images. The experiments were carried out by Zenil et al. (2012). We took seven images of the same format that we compressed using a lossless compression algorithm: the image recovered after decompression was exactly the same as the one before compression.

The first series indicates the ranking of images in ascending order of the file compression size. The classification is similar to that which the Kolmogorof complexity $K(s)$ would give. Unsurprisingly, the all-black image (first image in the top row) was found to have the smallest information content, and the image composed of randomly drawn pixels (last image) was the one requiring the greatest amount of memory (maximum $K(s)$). The other images were, as expected, classified approximately: the handwritten text (image placed in position two) required relatively little memory since there was a lot of white space; next was a periodic lattice (image three), followed by a Peano curve (image four), an irregular image with two axes of symmetry (image five) and, finally, a microprocessor (image six).

The second series of images (bottom line) employed the same seven images, but this time in increasing order of decompression time, which gave approximate values for Bennett's logical depth P(s) and therefore a classification by increasing structural complexity. Image seven in the first ranking, which was perfectly random, then became among the first, in keeping with the idea that perfect randomness is completely structureless. The microprocessor could be clearly identified as the most structurally complex. The Peano curve is unsurprisingly always considered to contain structures a little richer than the periodic pattern. The written text and the symmetrical pattern change position, resulting in a classification that is compatible with that which intuitively corresponds to increasing structural complexity.

Figure 3.1. *Results of experiments carried out on K(s) and P(s)*

3.10. References

Antunes, L. and Souto, A. (2010). Information measures for infinite sequences. *Theoretical Computer Science*, 411(26/28), 2602–2611.

Antunes, L., Matos, A., Souto, A., Vitanyi, P. (2009). Depth as randomness deficiency. *Theory of Computing Systems*, 45(4), 724–739.

Ay, N., Müller, M., Szkola, A. (2010). Effective complexity and its relation to logical depth. *IEEE Trans. Inform. Theory*, 56, 4593–4607.

Baptist, G. (2012). Cellule et futur biologie théorique et philosophie. PhD Thesis, Grenoble University, Grenoble.

Bennett, C. (1988). Logical depth and physical complexity. In *A Half-Century Survey on the Universal Turing Machine*, Herken, R. (ed.). Oxford University Press, New York.

Bennett, C. (2012). What increases when a self-organizing system organizes itself? Logical depth to the rescue [Online]. Available at: http://dabacon.org/pontiff/?p=5912.

Collier, J. (2008). Information in biological systems. *Handbook of Philosophy of Science*, 8, 763–787.

Danchin, A. (1998). *La barque de Delphes : ce que révèle le texte des génomes*. Odile Jacob, Paris.

Delahaye, J.-P. (1999). *Information, complexité et hasard*. Hermes, Paris.

Delahaye, J.-P. (2010). *Complexité aléatoire et complexité organisée*. Quae, Versailles.

Delahaye, J.-P. and Vidal, C. (2018). Organized complexity: Is big history a big computation? *American Philosophical Association Newsletter on Philosophy and Computers*, 17(2), 49–54.

Delahaye, J.-P. and Vidal, C. (2019). Universal ethics: Organized complexity as an intrinsic value. In *Evolution, Development and Complexity: Multiscale Evolutionary Models of Complex Adaptive Systems*, Yordanov, G.G., Martinez, C.F., Price, M.E., Smart, J.M. (eds). Springer, Berlin.

Gaucherel, C. (2014). Ecosystem complexity through the lens of logical depth: Capturing ecosystem individuality. *Biological Theory*, 9(4), 440–451.

Gell-Mann, M. (1995). *The Quark and the Jaguar: Adventures in the Simple and the Complex*. Macmillan, New York.

Grassberger, P. (2012). Randomness, information, and complexity [Online]. Available at: http://arxiv.org/abs/1208.3459.

Koppel, M. (1987). Complexity, depth, and sophistication. *Complexity*, 1, 1087–1091.

Li, L. and Vitányi, P. (2008). *An Introduction to Kolmogorov Complexity and Its Applications*. Springer, New York.

Mayfield, J. (2013). *The Engine of Complexity: Evolution as Computation*. Columbia University Press, New York.

Mitchell, M. (2009). *Complexity: A Guided Tour*. Oxford University Press, Oxford.

Soler-Toscano, F., Zenil, H., Delahaye, J.-P., Gauvrit, N. (2013). Correspondence and independence of numerical evaluations of algorithmic information measures. *Computability*, 2(2), 125–140.

Zenil, H., Delahaye, J.-P., Gaucherel, C. (2012). Image characterization and classification by physical complexity. *Complexity*, 17(3), 26–42.

4

Leveraging Complexity in Oncology – A Data Narrative

Where is the wisdom we have lost in knowledge?
Where is the knowledge we have lost in information?

T.S. ELLIOT

In this era when hypothesis-driven research is being superseded by the availability of huge datasets, information is organized in a critical process that shapes modern research. What data are available and how they can be retrieved provide the foundations of data-driven methods, which can be anathema for purists when the scientific method has been turned on its head.

A combination of data-gathering methods examining large datasets and hypothesis-driven approaches (validating identified leads deemed worthy of further study) might be the only way to understand complex diseases such as cancer. Needless to say, examining large datasets to reveal biomarkers correlating with a given condition will undoubtedly yield false positives. Ensuring the use of quality controls and robust statistical methods should lead to increased knowledge.

Biomedical research depends on the availability of and access to expensive technologies and infrastructures capable of handling large numbers of samples and the information generated by different analysis pipelines. The challenge we are

Chapter written by Xosé M. FERNÁNDEZ.

facing is not just circumscribed to the growing amount of data generated, but most importantly concerns its heterogeneity, as different technologies, or even platforms, will rely on substantially different (often incompatible) laboratory protocols. Further to this, data will end up spread all over data warehouses distributed across different hospitals and research centers.

Another key aspect to take into account is the context in which data are acquired, which is particularly relevant in clinical settings; temperature measurements taken during a routine examination will not have the same implications as thermometry measured in casualty departments, particularly if such data points are subsequently used to infer correlations.

For example, clinicians could use time of consultation as a proxy to understand survival, since blood tests requested in the middle of the night are often associated with traffic accidents, whilst blood tests ordered at noon are often associated with cholesterol or minor ailments. We could then erroneously find a correlation between survival and the time of day measurements were made, when the cause of death could be something completely unrelated, since emergency doctors routinely face extreme cases of casualties.

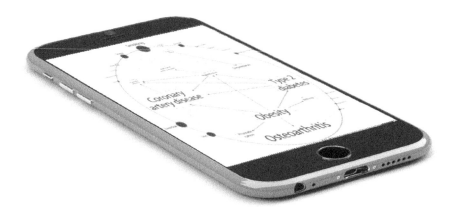

Figure 4.1. *Photomontage showing conditional dependency diagrams. Disease–environment interactions are shown with text and circles proportional to the number of diseases associated with each factor in the circuit. Modified from Ashely et al. (2010)*

Novel uses of data are not new to medicine, and there is historical evidence of new insights into disease revealed by data analysis. Let us look back at London in

August 1854, a city devastated by an outbreak of cholera[1]. John Snow was not new to cholera as he had witnessed, as a young surgeon's apprentice, the ravages caused by the disease in Newcastle in late 1831. Thanks to this previous experience he developed his waterborne theory, which would prove to be crucial during this Soho outbreak. His growing stockpile of data was presented in a striking map pinpointing the location of all fatalities during the outbreak; further improved versions incorporated additional data until the final version completed the visual impact by drawing a line labeled "boundary of equal distance between Broad Street pump and other pumps" in what later would be termed a Voronoi diagram. All this evidence made it crystal clear that cholera was radiating from that single pump, which eventually was disabled, putting an end to the episode. What a fine example of using mathematics to understand disease.

Data in physical sciences fit well-defined systems; however, when we turn to biological systems, we find multimodal data acquired even before there was a hypothesis, with datasets (medical records, genomic sequences, other clinical data, etc.) getting bigger and more complex. Data can be highly convoluted, given the significant dimensionality, nonlinearity and temporality present in most clinical contexts. In oncology, for instance, knowledge has been meticulously built over decades, primarily through carefully designed randomized, controlled clinical trials evolving longitudinally over the years and usually involving various potential biases pertaining to indirect comparisons.

Noisy, unstructured and dynamic data collections that may well be corrupted and incomplete, hiding an underlying global structure, are the ideal fabric for topological data analysis approaches. Networks as information representations are very amenable mathematical methods for the extraction of knowledge that can reveal communities, closely connected nodes as a group but with few connections outside. Recognizing communities within networks clarifies their structure, offering practical benefits; patients sharing similar treatments can be better stratified.

With such an explosion of datasets coming from big data initiatives, it is obvious that we not only require new tools but also a different mindset to accelerate research and practical innovations integrating multiple domains of information. Stochastic block models are a commonly used statistical method that assume the nodes within a community behave identically when interacting with other nodes. Such an approach may have application in forensics or drug selection from genetic data, particularly when using a wider family of node properties or a greater number of them.

---

1 No known case of cholera was reported in London before 1831.

4.1. Large collaborative research initiatives – the Human Genome Project

The initial working draft of the human genome published in 2001 (Lander et al. 2001; Venter et al. 2001) was the culmination of the scientific quest initiated at the beginning of the last century when Mendel's laws of heredity were rediscovered[2]. However, peering beyond the horizon of genomics, there is a rich landscape of exciting research endeavors: genome editing technologies, novel bioinformatics strategies and emerging international consortia, some in their infancy, but similar in ambition to the Human Genome Project (Collins et al. 2003).

The first draft of the human genome presented with great ceremony and ostentation in June 2000 generated numerous press releases, including unusual sources such as the White House announcing the *completion of an initial sequencing of the human genome,* which would usher us in to a *new era of molecular medicine.* According to U.S. President Bill Clinton, the outcome of the sequencing effort was "without a doubt (…) the most-wondrous map ever produced by humankind"[3]; in his words, humanity had exposed "the language in which God created life". Not only politicians but also Nobel laureate Jim Watson predicted that the impact of deciphering the genetic code "would revolutionize society as much as Gutenberg's printing press. After the printing press, there was an explosion; more people could have information. We'll understand ourselves better, have a better idea of what human nature is"; for Francis Collins (then leader of the public effort), it was "the most important and most significant project that humankind has ever mounted".

A draft encompassing 96% of the euchromatin and with an estimated 145,514 gaps was the first freeze of the human genome termed *hg1.* It was a first sketch to be improved (the version produced by the public endeavor) with subsequent updates scheduled (International Human Genome Sequencing Consortium 2004); the latest assembly (GRCh38), released in December 2013 with subsequent patches, provides a highly accurate sequence (99.999% accurate) but still features 349 gaps despite being the 20th iteration of the human genome made public (Schneider et al. 2017)[4]. Without questioning the validity of such a landmark endeavor, the current human reference genome introduces a pervasive reference allele bias as it considerably ignores common human variation. A single representative sequence originally

---

2 We should keep in mind that these laws were enunciated to understand the nature and content of genetic information.

3 As we will see in this chapter, alongside *language, code* and *book, map* is one of the most frequently invoked metaphors around genomics.

4 As this manuscript was going to press, the *Telomere-to-Telomere Consortium* published an improved human genome closing the remaining gaps persisting in GRCh38 and correcting numerous errors, but the Y chromosome is not fully sequenced so we should expect further updates (Aganezov et al. 2022).

produced for each chromosome has been expanded to accommodate representative alternate versions of highly variable regions with numerous "alternate locus scaffolds" stitched to the primary scaffolds (178 regions in the GRCh38 assembly).

However, when talking of the *human genome*, we could be misleading the unwary; one of the most constant facts we come across in our daily lives is variability among human beings. We find differences between individuals at the genomic level and this could make us wonder what was captured in the *reference* human genome. A reference summarizes, by definition, the huge diversity encoded in our genome, within the atomic structure of the DNA molecules. In this particular instance, we have a mosaic genome derived[5] from the sequences of 13 anonymous volunteers from different ethnic groups who happened to be in Buffalo[6], New York, in March 1997 (although most of the sequence, 71.92%, comes from one single anonymous male individual; the subsequent library was labelled RPCI-11). By the time this draft was published, only 3 million SNVs[7] were identified (0.2% of the genome), leaving the concept of race without any scientific basis (International HapMap Consortium 2007).

The molecular bases of almost one half of the 8,360 disorders with suspected Mendelian inheritance have been defined within the past generation (Manolio et al. 2009). Mendelian disorders account for approximately 17% of pediatric hospitalizations and an even greater proportion of healthcare costs.

4.2. Human cell atlas – unraveling complexity

Cells are the core unit of living organisms and essential to understanding the biology of health and the molecular events leading to disease. Single-cell sequencing should enable the detection of drug resistance and improve diagnosis, and so it will undoubtedly be pivotal for the future of cancer research. Researchers can now sequence single cells from biopsy samples, circulating tumor cells or surgical samples to find copy number variations enabling earlier diagnoses (Navin et al. 2011).

---

5 We focus on the efforts of the *International Human Genome Sequencing Consortium*; the private endeavour from *Celera Genomics Inc.* was based on five individuals, with the bulk of the sequence derived from J. Craig Venter (subsequently published; Levy et al. 2007).

6 Pieter J. de Jong from Roswell Park Cancer Institute advertised in *The Buffallo News* on March 23, 1997: "The Great Gene Hunt. Scientists Seek Man/Woman to Donate DNA".

7 In this contribution, we favor the use of single-nucleotide variants (SNVs) over the most widely used single-nucleotide polymorphisms (SNPs); as the latter include those instances only found in one particular individual, it could be said that the difference is a matter of frequency of occurrence.

Traditionally, map collections have been gathered in atlases, either general or thematic depending on whether they are devoted to a major geographical territory or theme. They brought forward the medieval way of thinking with sophisticated attempts to represent the planet as it was known at the time. Geographical precision was not the main purpose but a statement of ideology, articulating a contrast between the civilized world and the periphery. A compendium of the knowledge of the time was displayed in featuring cities and towns, pictures of animals and plants, alongside mythological creatures, with some reference to Biblical events.

When the human genome was first assembled at the turn of last century[8], our voyage charting genomes started and required the development of core biomolecular resources; there was a need to represent such information in a modern map; genome browsers brought those sequences to life (Cunningham et al. 2022; Lee et al. 2022).

We can track the evolutionary history of a tumor by mapping DNA variants in genes varying from person to person (Crosetto et al. 2017). Transcriptomics has revealed the complexity of multicellular organisms, pinpointing which genes are most active. However, we have applied these techniques to mashed tissues, losing many of the defining characteristics of individual cells along the way. We have yet to identify the different cellular lineages in humans and comprehensively catalog hundreds of cellular types and subtypes to create a high-resolution map of all the cell types in the human body.

Such an approach will easily require mapping and characterizing up to 1 billion cells throughout the human body's tissues, organs and systems so that we can also track the different states they go through in their lifetime, highlighting which genes are expressed (Tang et al. 2009) and which of the approximately 1,500 transcription factors (TFs) are being expressed at every point (Vaquerizas et al. 2009). This *dark matter* of living organisms adds an additional layer of complexity. For instance, we can find in the brain alone thousands of neuronal types, which express different sets of TFs and probably a larger variety of non-coding RNAs (Nelson et al. 2006).

Genes and genomes play a fundamental and even unique role in the functioning of life, but this role can only be understood when adequate attention is given to substances or structures interacting with DNA. This is how epigenetic processes (not requiring mutagenic changes to the nucleotide sequence) control the heritable transfer of information, so that cells with identical genomes can be differentiated

---

8 We have seen how scientists were engaged, during the 1990s, in a race to determine the complete DNA sequence in various model organisms: bacteria (*Haemophilus influenzae*), yeast (*Saccharomyces cerevisiae*), the round worm (*Caenorhabditis elegans*), the genetic workhorse *Drosophila melanogaster* and, as we have seen, the genetic blueprint of *Homo sapiens*.

into hundreds of distinct cell types within the human body (Alberts et al. 2008). Without this system of specialization allowing the coordination of gene activity and nuclear architecture, cells would be unable to correctly proceed through development and would fail to perform their vital functions. In order to maintain this lineage-specific identity, this mechanism must be able to be conserved through somatic replication.

Bioinformatics and experimental evidence suggest that we have misunderstood the genetic programming of complex organisms for the past 50 years. We have assumed that most genetic information is transacted by proteins, what is largely true for the unicellular prokaryotes but not for multicellular eukaryotes (Mattick 2010). A human genome programs the development in an individual of about 100 trillion cells with hundreds of different muscles, bones and organs, as well as the brain. It contains about 20,000 protein-coding genes, but we are finding pervasive transcription (ENCODE 2005, 2007) into ncRNAs, suggesting that much of the information required to program development may reside outside these sequences (Gerstein et al. 2007). In addition to ncRNAs, there are hundreds of thousands of RNAs showing specific expression patterns, a hidden network of regulatory information regulating epigenetic processes and directing the precise patterns of gene expression during growth and development in mammals. It also appears that RNA is central to brain development, learning and memory, and primates have developed sophisticated RNA editing systems to modify hardwired genetic information. RNA may represent the major substrate for gene–environment interactions in the cell. So what was dismissed as *junk* may hold the answer to understanding human evolution, development and cognition, as well as our idiosyncrasies and susceptibilities to complex diseases.

So-called big data are becoming pervasive in biomedical sciences and healthcare has ushered us into a world where data are growing faster than ever, both quantity and complexity posing new challenges. This is a revolution that touches every scientific discipline and every life on the planet.

4.3. From bench to bedside

We have seen how DNA is assumed to take the form of information with extensive use of metaphors of writing and programming; genomes can be read (gene sequencing) and written (see the latest developments in systems biology). Gilbert (1978) described the gene as *an expressed mosaic of sequences held in a matrix of silent DNA, an intronic matrix*, but there is more than protein coding. We will witness major changes in the way we approach personal health and the ability to monitor people as they go through their daily lives.

The availability of the human genome has paved the way to finding the connection between our genome and a growing number of diseases. Mapping the molecular pathways involved makes possible the development of diagnostic tools for new targeted therapies enabling clinicians to stratify patients by the molecular characteristics of clinical biomarkers (Leary et al. 2010).

Over the last decade, we have seen how the need to lower the cost of sequencing a human genome has led to the development of many massively parallel sequencing platforms. The field of massively parallel sequencing, more commonly known as next-generation sequencing (NGS), has advanced rapidly and has made it possible to analyze the genome at previously unfeasible depths (Schuster et al. 2008; Mardis 2009).

Publicly available genomes, thousands if not hundreds of thousands, will be required to understand the genetic bases of diseases and other phenotypes (we will certainly find rare variants at the core of most common diseases), but we are just taking the first steps in this long journey, which will unveil pathogenic mutations allowing us to begin to tease out function. Cancer, chronic diseases and rare diseases may need a million genomes[9], but this inconceivable figure can be achieved in a few years (Birney et al. 2018).

Precision medicine could well be defined as "a medical model using molecular profiling technologies for tailoring the right therapeutic strategy for the right person at the right time, and determine the predisposition to disease at the population level and to deliver timely and stratified prevention" (European Commission DR 2010).

The practice of medicine has the ultimate aspiration of being truly personalized. Any attempt to implement a personalized medicine strategy requires the confluence of several factors: suitable IT infrastructure for the healthcare system (i.e. capable of handling electronic medical records (EMRs)), regulation (e.g. the General Data Protection Regulation, (EU) 2016/679, approved by the European Parliament and the Council of the European Union in 2016 and enacted to bring European law into line with the latest technical developments (95/46/EC)), technology and tools, insurance coverage (universal in Europe), genetic privacy and legal frameworks (e.g. H.R.493: Genetic Information Nondiscrimination Act of 2008 and European legislation) and medical education[10]:

---

9 As people seek better health outcomes, it is estimated that about 20 exabytes of genomic data should be available.

10 According to a survey led by the HuGEN Project, the majority of medical practitioners (70%) had discussed genetics with their patients and 30% reported providing counseling about genetic

Unless Americans are convinced that the information will not be used against them, the era of personalized medicine will never come to pass (Collins 2007).

Sequencing is not the end of the game, particularly when genomics is transitioning into mainstream clinics as ambitious national initiatives attempt to deliver significant health benefits by introducing genomics into their health systems (we could name here *France Médecine Génomique 2025* or *The 100,000 Genomes Project* from *Genomics England*): sample collection and tracking, analysis and interpretation, secure storage and exploitation are amongst a new set of challenges presented.

Ensuring the success of precision medicine will require a continually updated longitudinal life-course health record from primary care and hospitals all the way to outcomes. An early and interesting insight into this new era was provided by the analysis of Stephen Quake's genome[11] (Pushkarev et al. 2009). Several months after the publication of his genome, a clinical assessment was published (Ashley et al. 2010). The proband has a family history of vascular disease and early sudden death. Note that 2.6 million SNVs and 752 copy number alterations (CNAs) were analyzed to shed some light on increased genetic risk for myocardial infarction, type 2 diabetes and some cancers. There were also a number of pharmacogenetic markers allowing some prediction of his response to different medications in the future: clopidogrel resistance associated with a loss-of-function mutation in *CYP2C19*, low dosing requirement for warfarin associated with *VKORC1* and *CYP4F2* mutations, etc. Following this analysis, the mutations associated with cardiomyopathies prompted him to have an echocardiogram. More than ever, the ability to manage torrents of data is critical to our success in taking on the most challenging problems in health and medicine, as we rely on computing expertise in order to find information in the midst of vast datasets.

4.4. The battle with cancer

Cancer is not a new disease; it is the evolutionary legacy toll multicellular organisms pay (because this ailment is not restricted to humans and we can trace

---

concerns. However, fewer than 10% felt confident in their training in this matter (Lapham et al. 2000).

11 Stephen Quake, a Stanford professor, sequenced his own genome, publishing it as a showcase for a new sequencing platform of his invention (Pushkarev et al. 2009). It only took a single machine (worth $1 million) and extensive sequencing (costing less than $48,000) to shatter the promised dawn of a new age (bringing the use of these technologies closer to the clinic). A year later, a paper with 31 authors (whereas the original sequencing paper only had three authors) provided a clinical assessment showing why this is not common practice yet. It is not trivial making sense of the sequence (Ashley et al. 2010).

early signs in the fossil record and mummies[12]). It could be defined as an information disease (all cancers have a genetic base) that results from the proliferation of tumor cells invading neighboring parts of the body. Thus, cancer is not one emperor of maladies but a whole empire of related conditions spreading throughout the human body; an invincible grim reaper enshrined in its own mythology, as Susan Sontag presented in her *Illness as Metaphor*.

For thousands of years, there was only one treatment available for cancer, that is, surgery. In 1896[13], a second pillar, radiotherapy, was added (Despeignes 1896). The foundations for the third treatment pillar (cancer chemotherapy) were laid by Gustaf Lindskog in 1943 with the first use of nitrogen mustard as a treatment for non-Hodgkin's lymphoma (Goodman et al. 1946). These three pillars – surgery, radiotherapy and chemotherapy – continued to provide the backbone of cancer care until precision therapeutics (the majority of the 70 new anticancer compounds approved in the last 5 years, as well as the bulk of the 1,800 molecules in the pipeline, belong to this category) and soon afterwards immunotherapy (with dozens of immunotherapy agents approved in the last few years) were added to the weaponry[14] available against cancer.

This therapeutic revolution is shaking the foundations of cancer treatment. Immunotherapy has reversed a trend toward more ambulatory approaches to cancer treatments, and cancer is no longer treated as an uncontrolled cellular degeneration but also as a failure of the immune system. A more personalized approach to medicine focuses now on the immune system of the host rather than just characterizing the tumor (Harrington et al. 2005). An increasing number of targeted inhibitors and their use driven by the molecular landscape of the tumor, either individually or in combination with other therapies, are giving hope for new breakthroughs in cancer treatment.

---

12 The oldest descriptions of cancer in humans can be found in Egyptian papyri from ca. 1550–1825 BCE, which list conditions consistent with modern description of cancer. Even older skeletal remains, skulls and femurs from the Bronze Age show structures compatible with primary bone tumors or lesions suggestive of metastasized cancer (Lieverse et al. 2014).

13 François-Victor Despeignes (1866–1937) was possibly the first person to use X-rays to treat cancer when he treated a 52 year old male patient most likely with a lymphoma (although he reported a *cancer de l'estomac* (Laugier 1996)) just 5 months after Röntgen discovered X-rays.

14 President Nixon declared a "War on Cancer" back in 1971, and often some variant of the term "battling cancer" is used in describing all sorts of oncologic treatment, from undergoing locoregional therapies such as surgery and/or radiation therapy to seeking systemic chemotherapies or biologic therapies and even pursuing alternative regimens.

Precision medicine[15] promises a paradigm shift in care delivery, one that removes the need for guesswork and treatment strategies based on generalized demographics. The huge genetic diversity within a tumor driven by a range of genetic mutations is currently being catalogued in order to allow more targeted, potentially more effective therapies, which will have better patient outcomes.

Oncology has arguably led the development of precision medicine in recent decades, with genomic and molecular advances informing the development of targeted therapies that have revolutionized treatment of certain individual cancer types. Today, we can offer a genomic or "molecular" diagnosis, which means that we better understand the genetic base and can use this information to help select the most effective treatment, greatly improving chances of survival. This can be used for a wide range of cancers, such as melanoma (skin cancer), leukemia and colon, brain and breast cancers. This understanding means that cancer patients can be stratified according to what will be most effective for their condition, and it may also mean that patients with different types of cancer may, on the basis of the genomic diagnosis, receive similar treatments.

Cancer has commonly been described as a disease of the tissue of origin (e.g. breast cancer, colorectal cancer, etc.) and its cell types (sarcoma, carcinoma, etc.) However, in the biological realm, the implementation of microarray platforms facilitated the identification of several clinically useful gene expression signatures, bringing improvements into clinical management. A more recent wave of whole-genome (and exome) sequencing analyses has revealed an underlying wealth of tumor-specific (somatic) alterations that can be used to improve tumor classification. Those signatures of biomarkers are facilitating moving patient stratification beyond the research arena to clinical care, and there is good evidence to suggest that each of these subtypes may exhibit a predictable clinical phenotype.

Blood cancers provide a good example. In the 1920s, there were only two diagnoses available: leukemia and lymphoma. Today, cellular and genetic analyses can distinguish 38 types of leukemia and 51 types of lymphoma. A better understanding of the nuanced differences between the categories of cancer has aided drug development in identifying targeted treatment for several of these cancer subtypes, with the resulting increase in survival rates from 0% to as high as 90% in some cases.

---

15 Currently, there are about 800 cancer drugs in development, many of them specifically designed to target a given set of mutations. It may take changes in regulatory policies, as we can already see with the latest approvals from the EMA and FDA (e.g. erlotinib), and the development of new diagnostic tests in order for successful therapies to make it to the market (termed *companion diagnostics*).

4.5. Health economics – cost is another matter

Perhaps the current Cancer Moonshot is not a novel initiative, as there has been a demand for years; we can witness in *The New York Times* (August 5, 2009) that Jim Watson decried the lack of leadership, recollecting the war on cancer declared by US President Richard Nixon in 1971[16] and the *National Cancer Act* passed by US Congress in December that year (the NCI's budget would quadruple over the following five years):

> The National Cancer Institute, which has overseen American efforts on researching and combating cancers since 1971, should take on an ambitious new goal for the next decade: the development of new drugs that will provide lifelong cures for many, if not all, major cancers. Beating cancer now is a realistic ambition because, at long last, we largely know its true genetic and chemical characteristics.

The Nobel laureate calls out for biochemists to focus their research in the field of the cellular chemistry of cancer cells.

> After my colleagues and I discovered the double helix of DNA, biology's top dogs then became its molecular biologists, whose primary role was finding out how the information encoded by DNA sequences was used to make the nucleic acid and protein components of cells. Clever biochemists must again come to the fore to help us understand the cancer cell chemically as well as we do genetically.

Nowadays, without mounting popular pressure in the media[17], we are witnessing how the model pioneered by the Human Genome Project Consortium has allowed the establishment of new international efforts in genomics. One example is the International Cancer Genome Consortium (ICGC 2010), which opened new ground; an ambitious plan to sequence thousands of cancer genomes to redefine the landscape in oncology and currently extended as the International Cancer Genome Consortium Accelerating Research in Genomic Oncology (ICGC ARGO) to incorporate clinical information (treatment, response, outcomes and any other health data available).

Understanding how cancer arises requires more than a list identifying which genes are mutated in certain cancers, although this is a good start. We are witnessing

---

16 Public pressure was increasing as new antibiotics were added to the clinician's toolset (chloramphenicol, tetracycline, streptomycin) following in the footsteps of penicillin, but there was still no magic bullet available to cure this dreadful disease; cancer remained a black box.

17 Some opponents of the Human Genome Project have recently voiced their criticisms with the celebration of the 10th anniversary of the publication of the first draft.

a paradigm shift where cancers are classified by their mutations, not by affected organ or tissue (there are already innovative molecularly targeted clinical trials, *basket trials*, based on tumor molecular profiling (Le Tourneau et al. 2015)). *BRCA* mutations can be found beyond breast and ovarian cancers, but also in prostate and pancreatic cancers, providing new metabolic pathways to target with new drugs (Mateo et al. 2015; Valsecchi et al. 2014). Unfortunately, genetic tests are not yet routinely used in clinical practice (even those targeting mutations known for years). In conditions such as lung cancer, the number of actionable genetic mutations that can determine which therapeutic strategy is most likely to work is increasing; in non-small-cell lung cancer (NSCLC) alone, genetic testing with *EGFR*[18] can predict whether the patient will respond to a certain class of drugs (e.g. gefitinib), but less than 25% of patients get tested[19] (Chmielecki et al. 2011). Another drug, crizotinib, has been approved for treatment of some non-small-cell lung carcinomas; it is an *ALK* inhibitor (4% of patients with NSCLC have a chromosomal rearrangement that generates a fusion gene between *EML4* and *ALK*).

Figure 4.2. *US Food and Drug Administration approvals provided new hope to patients diagnosed with lung cancer; in only 12 months, there has been the approval of two immunotherapeutics for NSCLC (pembrolizumab and novolumab), an array of molecularly targeted therapeutics (osimertinib, necitumumab, crizotinib and alectinib), the expansion of crizotinib and the approval of the cobas EGFR Mutation Test v2 for liquid biopsies in June 2016*

---

18 The epidermal growth factor receptor (EGFR) is the cell-surface receptor for members of the EGF family of extracellular protein ligands. EGFR is a member of the ErbB family of receptors.

19 This is not the case in France, where the *Institute National du Cancer* is fostering molecular medicine following a decision taken in 2005 "to pay for the treatment of every citizen shown to be likely to benefit from targeted drugs". Biopsies from cancer patients are tested with 20 genetic tests and if the tissue displays a genetic signature in any molecular pathway targeted by one of the drugs, the patient gets treated with it. This approach is cost-effective: €35 million for 1,700 patients treated until they stopped responding (~38 weeks) instead of €104 million for all 15,000 patients treated for 8 weeks to see whether they would respond or not.

The Cancer Genome Atlas (TCGA) coordinated by the NCI and NHGRI to catalog and discover major cancer-causing genomic changes was expanded as part of the ICGC following the initial success of an effort led by the NIH to map genomic changes in glioblastoma multiforme, ovarian and lung cancers.

Tim Ley and colleagues published the first whole-genome sequenced tumor, a type of leukemia called AML (Ley et al. 2008). In this seminal publication, the researchers identified 10 mutated genes (2 mutations previously described in $FLT3^{ITD}$ and $NPM1^{CATGins}$ genes) and 8 new mutations present in virtually all tumor cells at presentation and relapse ($CDH24^{Y590X}$, $PCLKC^{P1004L}$, $GPR123^{T38I}$, $EBI2^{A338V}$, $PTPRT^{P1235L}$, $KNDC1^{L799F}$, $SLC15A1^{W77X}$ and $GRINL1B^{R176H}$). A method for discovering cancer-initiating mutations in previously unidentified genes may respond to targeted therapies.

Without attempting to review all the different cancers which have been sequenced to date (which would be beyond this contribution), we can note that targeted treatments for cancer have been extending lives for almost 20 years. It all started when trastuzumab (Herceptin®), a monoclonal antibody against the Erbb2 receptor, was approved in 1998 (the first biosimilars are about to hit the market). Most melanomas (~60%) show a gain-of-function mutation involving codon 600 of the $BRAF$ oncogen[20] ($BRAF^{V600E}$), also recurrent in colorectal (10%) and thyroid cancers. Vemurafenib was approved in August 2011, along with a companion diagnostic kit for the treatment of metastatic melanoma in patients with $BRAF^{V600E}$. Dramatic responses to Vemurafenib have been observed in patients with metastatic melanoma expressing $BRAF$ (81% response rate and extended overall survival by 7 months). However, this is a transient success as patients relapse after acquiring specific resistance to the drug within a few months.

Immunotherapy is one the most exciting developments in cancer treatment in years, beginning to take off at a time when much cancer drug research seemed to be dwindling. The effects of more orthodox chemotherapy drugs were hitting a wall[21], extending the life of patients a few months or even just a few weeks. However, the cost of cancer treatments nowadays exceeds €7000 per month over periods of up to 6 months and the bill for immunotherapies goes even higher (prices exceeding $150,000

---

20 The B-Raf protein is in the chain of proteins known as the mitogen-activated protein kinase (MAPK) pathway. MAPK relays growth signals from outside the cells to the nucleus.

21 France spends €1.4 billion per year in chemotherapy drugs, and this figure will progressively increase by €200 million each year.

per infusion) while CAR-T[22] cell therapy may cost a staggering $850,000 per patient[23]. It is quite difficult to see how this would be sustainable for our aging populations.

4.6. From molecules to medicine

We are living through a sustained renaissance of medicine that started with the initial sequencing of the human genome. The power of genomics lies in its objective ability to correlate with physical manifestations in the patient so we can harness the effect of these phenotypes. Precision medicine enables us to quantify human health and disease, a fundamentally different approach to healthcare where the focus is on identifying and addressing individual patient differences rather than making broader generalizations.

The practice of medicine involves screening the general population and diagnosing suspected cases before intervention on a given patient. Precision medicine has not focused yet on screening; before we get there, we require extensive data collection and analysis revealing genomic patterns in the molecular phenotype of the patient. We can then deploy powerful big data algorithms for population screening of healthy individuals through digital tools fueled by that knowledge.

Despite advances in pharmacology, various harmful and adverse effects are still common. Some patients will not benefit from their treatment or experience an adverse reaction to the medication. Progress in the understanding of disease mechanisms and drug actions is opening opportunities to match therapies to patient subgroups and thus pave the way toward a more personalized medicine. Molecular markers are the cornerstone of new classifications of disease. New drug targets can be identified based on this molecular profiling of disease.

Adverse drug reactions (ADRs) have reached epidemic proportions and are increasing. The European Commission estimated in 2008 that adverse reactions kill 197,000 EU citizens annually at a cost of €79 billion; figures in the United States are not different, with 700,000 A&E visits and 120,000 hospitalizations annually due to ADRs and $3.5 billion spent on extra medical costs of ADRs per year.

Nowadays, the tools provided to the medical practitioner should be more precise, going beyond the obvious such as a tumor on a mammogram or the histological

---

22 Chimaeric antigen receptors are a class of synthetic receptors that reprogram immune cells for therapeutic purposes. Genetically engineered T cells (comprising three canonical domains for antigen recognition, T-cell activation and costimulation) are offering hope for curative responses in cancer.

23 This is the initial estimated cost of a one-time treatment with Luxturna gene therapy against a rare, inherited retinal disease that can lead to blindness ($425,000 per eye).

grading to the molecular profile of each individual. We have witnessed some developments in this domain with computer systems capable of answering questions posed in natural language (cohort profiling tools or patient browsers such as *Continuum Soins Recherche* (Consore)[24] employ complex queries and aim to improve clinical trial enrolment (Heudel et al. 2016)). The final goal of such research projects is the development of a clinical decision support system to aid the diagnoses and treatment of patients. Such stratification relies on a deep understanding of the underlying molecular mechanisms involved in the different pathologies.

Innovation is not easy in any industry, and producing new drugs shows just how hard this can be. Manufacturing drugs is cheap, but what is extremely expensive is to develop new drugs[25]. The pharmaceutical industry spent $59 billion in 2015 to put 45 new drugs in the market. The major reason for the rising cost of new drugs is the fact that more than 90% of them fail in clinical trials. In the period from 1950 to 2008, the Food and Drug Administration (FDA) approved 1,222 new drugs[26], and despite the fact that total R&D expenditure by governments and industry has tripled (in real terms) since 1999, the number of new molecular entities[27] (NMEs) approved by the FDA (similarly in Europe with the European Medicines Agency (EMA)) to be used as drugs has been in decline[28]. Moreover, there are an increasing number of NMEs approved with a "warning" label (almost half of them, 47%), indicating they are effective in treating only subpopulations of people with a given disease. Meanwhile, submissions for regulatory approval of new drugs and therapeutic indications are dwindling.

---

24 Consore and other tools deployed at the Institut Curie rely on natural language processing (NLP) approaches to read doctors' notes, pathology reports, diagnoses and recommendations and to detect hard-to-find real world data. They then graph this information and match appropriate patients with search criteria, potentially reducing recruitment time for clinical trials.

25 Note that $1,160 billion was the estimated cost of cancer in 2010, and this cost is increasing at regular intervals.

26 In 1950, more than half the medicines in common medical use were unknown a decade earlier (Smith et al. 1948).

27 A new molecular entity is any medication containing an active ingredient that has not been previously approved for marketing in any form by the FDA in the United States. The term "NME" is conventionally used to refer only to small-molecule drugs, but in this instance the term includes biologics as a shorthand for both types of new drugs.

28 A number declining from an average of 33 NMEs annually approved during the period 1993–1997 to 23.5 during 2000–2009, but a successful year (2017) reverted this trend, bringing the average from 2010 to 33 NMEs (and these figures do not include gene therapies and CAR-T). The industry has responded with pharma increasing R&D spending by 160% (from $15 billion to $37 billion from 1995 to 2005) and similar increases can be seen in the biotech industry, with a 150% increase (from $8 billion to $20 billion) in R&D spending during the same period.

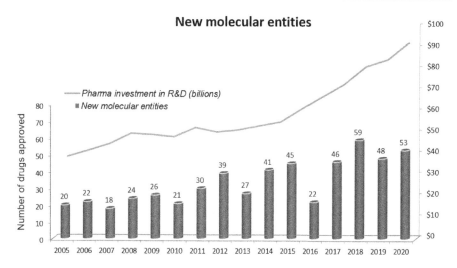

Figure 4.3. *NMEs approved by the FDA in the period 2005–2020 alongside investment in R&D from the pharmaceutical industry. Source: FDA and Pharmaceutical Research and Manufacturers of America*

Amongst the barriers hindering the evolution from *classic* trial-and-error medicine to a stratified approach (precision medicine), we can find the pharmaceutical industry. An obsolete *blockbuster model* centered on developing and marketing drugs for broad patient groups discourages aiming at smaller subpopulations. With the current model, precision medicine would appear to go against profitability; however, as we can see with imatinib, a tyrosine kinase inhibitor for chronic myeloid leukemia has transformed this ailment into a chronic disease, with patients taking the drug Gleevec® for life. Similarly, the regulatory environment requires investing too many resources in phase III clinical trials (addressing efficacy and safety) but too few in monitoring and assessment after FDA approval. This *fire and forget* model is about to change with patient-reported outcomes where cancer patients can provide feedback on their treatments.

4.7. Artificial intelligence

Human beings strive to augment their abilities by building tools, and this remains unchanged from the early stone toolmaking to the Hubble telescope. Within just the past few years, machine learning has become far more effective and widely available, and data scientists are eager to take on the most challenging problems in medicine and will be developing new tools to effect this change.

Every second, billions of transistors in billions of computers switch billions of times. Machine learning is taking automation to new heights[29], and this will bring extensive economic and social changes, as the Internet, the personal computer, the automobile and the steam engine did in their time. Humans have been using these technological innovations to adapt the world to us rather than adapting ourselves to the world, our digital evolution. Machine learning is a new chapter in this million-year saga and the revolution in biomedical research with ripples into medicine has been catalyzed by artificial intelligence and machine learning approaches.

The concept of an "intelligence explosion" (Good 1965) has been distorted in a science-fiction narrative but we could argue that it was flawed from the beginning in considering intelligence (like many other early theories about artificial intelligence (AI)) as a decontextualized abstract concept. AI is not a new sort of magical alchemy or a black box that simply ingests mountains of data and then just produces results; it is an enormous set of technologies, each with a specific, fine-tuned purpose.

Machine learning methods try to simulate human thinking in broad terms[30]; the machine learns from examples and can detect trends and outcomes in clinical trials without being explicitly programmed for a particular outcome. Alongside this application, it is rapidly being adopted with great impact in other biomedical applications, such as image, text and speech processing; however, it requires a different software approach (not that different to the way of writing code for cloud-based systems in opposition to code run in high-performance computing environments, parallelizing processes in the latter and run on many nodes to reduce costs).

AI is the advanced stage[31], adding enormous potential in its ability to recognize patterns and draw inferences from large volumes of EMRs, medical images, epidemiological series and other data. AI combined with healthcare digitization can allow us to monitor patients remotely (real world evidence (RWE)). AI is not intended to replace but to augment and assist human intelligence, incorporate user-

---

29 Machine learning encompasses the application of any computer-enabled algorithm to a dataset to find a pattern in the data and keep improving its performance without humans having to explain exactly how to accomplish all the tasks it is given. This can involve either supervised, unsupervised, segmentation, classification or regression algorithms.

30 Logistical regression and "old-fashioned statistics" are simple machine learning examples; iterative techniques are often used in trial designs.

31 The term *artificial intelligence* was coined in 1955 by John McCarthy, who organized the first seminal Dartmouth Artificial Intelligence conference with Claude Shannon and Martin Minsky in 1956. Ever since, perhaps in part because of its suggestive name, the discipline has given rise to more than its share of fantastic promises and assertions.

friendly efficiency, leverage data and make predictions. At Institut Curie, we are very active in this field, with a number of projects and partnerships around AI-based image recognition and working out how machine learning approaches for MRI and mammograms can help our radiologists[32].

All the recent advances in this realm are the result of a class of algorithms collectively known as *deep learning*, which allow higher levels of abstraction and improved predictions from data as it consists of a hierarchy of layers where each layer transforms the input data into more abstract representations (LeCun 2015). They can make better use of much larger datasets than previous approaches as they can extract rules and patterns from datasets, overcoming the classical plateau after which additional data did not improve predictions even if the number of examples in the training dataset increased substantially (Schmidhuber 2015).

The expertise of accomplished clinicians has been dissected numerous times, revealing that clinical decisions involve more than science; capturing the "art" of a skilled clinician is proving a very elusive challenge. A universal finding is that physicians respond to cues in the clinical data by conceptualizing some key diagnostic tasks central to the decision-making process (i.e. subtle hints in a patient record). The role played by this conceptualization governs any further data acquisition or alternatives brought into consideration in the decision-making process (additional dimensions such as the reliability of the resource can be considered when assessing the value of information). Decision trees are widely used in medicine and can be represented by a simple neural network with only two hidden layers (all the decision nodes of the tree will be on the first hidden layer and all the leaf nodes on the second hidden layer). Deeper trees will get wider (and not deeper) networks (Sethi 1990).

Our brain contains roughly 86 billion neurons; thus, any decision we make could be considered the outcome of neural computation[33]. Artificial neural network algorithms have recently advanced to detecting patterns without the prior definition of features or characteristics because of distributed parallel systems using graphics processing units (GPUs) and field-programmable gate array (FPGA) integrated circuits on large networks of processors. Originally, such decision tree-like applications were built manually based on expert human input, while nowadays they

---

32 Preliminary results indicate that the generic descriptors extracted from CNNs are extremely effective in object recognition and localization with natural images.

33 Frank Rosenblatt improved learning methods by introducing a training algorithm providing one of the first procedures for supervised learning of binary classifiers, the *perceptron* (Rosenblatt 1958).

can be evolved using multiobjective decision tree algorithms[34], a perfect combination in clinical settings.

Having two adversarial networks learning different features in parallel for unsupervised domain adaptation can help us pinpointing when things go wrong. We can have a primary network learning about tumor segmentation from scanned images, and a second network looking at the learnt representations to figure out the data origin from the convolutional neural network (in order to avoid bias based on the test center or even platform differences as scanners include new features).

Cognitive computing will be the next level beyond AI, as it takes into account the interplay of disparate components and systems to learn at scale, reason with purpose and interact with humans naturally. In the biomedical domain, this includes combining multimodal data from disperse sources (clinical data, EMRs, medical imaging data and the medical literature) in order to create more precise and cost-effective medicine. Radiologists are challenged with ever increasing amounts of clinical data and images from different modalities that they need to process in reduced time slots. Cognitive computing machines would complete the initial data analysis, filtering out the obvious cases and leaving for the radiologists the more complex cases with easy access to all multimodal information.

As the data deluge continues, we are finding newer ways of managing and analyzing to gather actionable insights and grapple with the challenges of security and privacy. Big data processes codify the past. They do not invent the future, which would require imagination, and that is something only humans can provide.

4.8. The fourth paradigm

Life's endless diversity is the result of a single mechanism: natural selection. Such an achievement is even more remarkable as it relies on a mechanism very familiar to computer scientists: iterative search, where a problem is solved by trying many alternative solutions, selecting and modifying the best ones, and repeating all these steps as many times as required. Evolution is the ultimate example of how much can be achieved with a simple learning algorithm as long as we are given

---

34 Neural networks are not a new idea, we can trace the concept back to the 1940s with the first machine learning architectures inspired by the structure and function of the human brain. Warren McCulloch and Walter Pitts proposed a model of artificial networks in which each neuron was postulated as being in a binary state (using what came to be known as a Turing machine) to understand how the brain could produce highly complex patterns relying on neuronal networks (McCulloch and Pitt, 1943). Surprisingly, this theory was not adopted by neurobiologists.

enough data, but data are only useful when accessible; therefore, suitable infrastructures are required to support this access.

Information and communication technologies are merging with biomedical sciences to create a new capacity to acquire, store, distribute, match and interpret vast volumes of complex data from patients, research, clinical trials and population health studies. They also have exciting potential to fundamentally change our concept of health and medicine and positively drive advances in healthcare.

It is important to distinguish between information and data: information, as Peter Drucker defined it, is "data endowed with relevance and purpose". More than ever, the ability to manage torrents of data is critical[35] as they will be of limited value until they have been integrated with other data and transformed into information that can guide decision making. Data architecture describes how data are collected, stored, transformed, distributed and consumed. It includes the rules governing structured formats, such as databases and file systems, and the systems for connecting data with the processes that consume it (on average, less than half of an organization's structured data are actively used, and less than 1% of its unstructured data are analyzed or used at all). Information architecture governs the processes and rules that convert data into useful information; these "plumbing" aspects of data management are vital to high performance (it is estimated that hospitals produced more than 665 terabytes of data in 2016).

4.9. Modeling the complexity of cancer

Complexity is the defining trait of science in the 21st century. Biology connected with engineering in the 1970s, producing high-throughput biology, and the 1990s brought us the Human Genome Project with cross-disciplinary biology. With ever increasing amounts of patient data produced, currently we are seeking systems capable of handling 14 million new cancers annually (about one genome every two seconds)[36], which will rely on elastic scalability and robust resource orchestration, the dream of numerous cloud providers.

Genomics research is complex and an understanding of its medical potential and the ethical issues involved requires a basic understanding of the principles of genetics. This will only be achieved

---

35 It is estimated that 55% of the data centers needed to process data around the world in 2025 have not yet been built, and we are building our new data center in Saint-Cloud (western Paris).
36 Every year, around 3.2 million Europeans are diagnosed with cancer (1,053 new cases every day in France), which is also the most common cause of death in France and most developed countries in Europe.

by a major effort to increase the quality of education in this field at all levels, with particular emphasis on improving science teaching and the introduction of the principles of ethics to school children.[37]

Medical advances in precision medicine face substantial challenges. Biomedicine has been lagging behind technologically for decades but we have seen how it has now embraced new technologies (high-throughput sequencing, wearable sensors, improved medical imaging, etc.), which, alongside advances in software pipelines, can handle vast amounts of information. However, managing data collection and data interpretation are still the biggest challenges, alongside safe storage and long-term preservation.

We have an unprecedented opportunity to deliver translational medicine. In other words, we are aiming to exploit the newly found knowledge from integrative efforts that should be disseminated to biomedical scientists, clinicians and patients. This requires the development of analytic, storage and interpretive methods to produce novel techniques for the integration of biological and clinical data.

Voluminous genomic and biological information should be integrated into diagnostic tools for the clinician. Predictions of science are limited to what we can systematically observe and model. Genome browsers paved the way to genomic research (Nielsen et al. 2010), and we have seen how genomics is changing the way we study cancer (Mardis and Wilson 2009; Pleasance et al. 2010). Tumors are very unpredictable systems, with mutant cells competing for space and resources, evading the immune system attacks; any model would be mimicking natural evolution in a tumor microenvironment heaving with different cell populations (stromal, immune, endothelial, etc.).

Historically, it was considered that cancer proliferated from a unique cell, and all cancer cells were the same, when in reality every single cell accumulates mutations, which surprisingly are usually well-tolerated by the cells in our body. A small handful of mutations are needed for a normal cell to turn into a cancer cell (Martincorena et al. 2017). Solid tumors are much more complex than an isolated mass of proliferating cancer cells. Mathematical models have been developed to address questions such as tumor initiation, progression and metastases, as well as intra-tumor heterogeneity, treatment responses and resistance; the complexity of cancer is well-suited to such quantitative approaches. Chaotic cancer models have been used to model tumor dynamics where we need to take into account not only the host, but also immune and tumor cell proliferation as key components of the tumor microenvironment.

---

37 WHO (2002). Genomics and World Health. Geneva.

$$\dot{x} = \rho_1 x(1-x) - \alpha_{13}xz$$

$$\dot{y} = \frac{\rho_2 yz}{1+z} - \alpha_{23}yz - \delta_2 y,$$

$$\dot{z} = \rho_3 z(1-z) - \alpha_{31}xz + \alpha_{32}yz.$$

This quantitative model (Letellier et al. 2013) describes the interactions between the host x, immune cell populations y and tumor cell populations z competing in a tumor (with a growing rate ρ_1 and inhibiting host cells at the rate α_{13}); the first term in the second equation represents the immune response to tumor antigens (with effector immune cells inactivated by tumor cells at the rate α_{23} and naturally dying at the rate δ_2). The final equation provides the growth rate of tumor cells, where the first term is a logistic function with governs the tumor cells when they are alone (with a growth rate of ρ_3). Competition between host cells and tumor cells is described by a degradation of tumor cells according to α_{31}. Tumor cells are killed by effector immune cells at the rate α_{32}. This model could be extended with additional equations; for example, endothelial cells. The model above could be simplified according to Itik and Banks, who set $\rho3$ and $\alpha31$ to 1, defining a model in competition, which can be seen in the fluence graph in Figure 4.4.

Box 4.1. *Quantitative model*

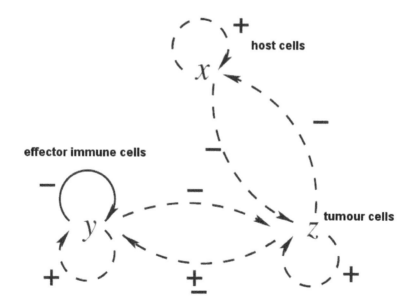

Figure 4.4. *Fluence graph for the cancer model described in the text. Solid lines represent linear interactions and dashed lines nonlinear interactions promoting (+) or repressing (–) the growth of the cellular lineage*

Cancer models complement experimental and clinical studies not only by challenging the current paradigm but also by elucidating mechanisms driving tumorigenesis and validating quantitative predictions. The power of mathematical modeling lies in its ability to reveal previously unknown principles and as more quantitative measurements become available from preclinical models and patient samples, it will become more important to improve our understanding of cancer biology and treatment.

4.10. References

Aganezov, S., Yan, S.M., Soto, D.C., Kirsche, M., Zarate, S., Avdeyev, P., Taylor, D.J. et al. (2022). A complete reference genome improves analysis of human genetic variation. *Science*, 376(6588), eabl3533.

Alberts, B.A., Johnson, A., Lewis, J., Raff, M., Roberts, K., Walter, P. (2008). *Molecular Biology of the Cell*. Garland Science, London.

Ashley, E.A., Butte, A.J., Wheeler, M.T., Chen, R., Klein, T.E., Dewey, F.E., Dudley, J.T. et al. (2010). Clinical assessment incorporating a personal genome. *Lancet*, 375(9725), 1525–1535.

Birney, E., Vamathevan, J., Goodhand, P. (2022). Genomics in healthcare: GA4GH looks to 2022. *Biorxiv* [Online]. Available at: https://doi.org/10.1101/203554.

Chmielecki, J., Foo, J., Oxnard, G.R., Hutchinson, K., Ohashi, K., Somwar, R., Wang, L. et al. (2011). Optimization of dosing for EGFR-mutant non-small cell lung cancer with evolutionary cancer modeling. *Science Translational Medicine*, 3(90), 90ra59.

Collins, F.S. (2007). The threat of genetic discrimination to the promise of personalized medicine. Genetic Information Nondiscrimination Act of 2007. Testimony, HR 493, Subcommittee on Health/Committee on Energy and Commerce/United States House of Representatives Hearing.

Collins, F.S., Morgan, M., Patrinos. A. (2003). The Human Genome Project: Lessons from large-scale biology. *Science*, 300(5617), 286–290.

Crosetto, N., Bienko, M., van Oudenaarden, A. (2015). Spatially resolved transcriptomics and beyond. *Nature Reviews Genetics*, 16, 57–66.

Cunningham, F., Allen, J.E., Allen, J., Alvarez-Jarreta, J., Ridwan Amode, M., Armean, I.M., Austine-Orimoloye, O. et al. (2022). Ensembl 2022. *Nucleic Acids Research*, 50(D1), D988–D995.

Despeignes, V. (1896). Observation concernant un cas de cancer de l'estomac traité par les rayons Röntgen. *Lyon Méd.*, 428–430.

Dolgin, E. (2009). Collins sets out his vision for the NIH. *Nature*, 460(7258), 939.

Drucker, P.F., Collins, J., Kotler, P., Kouzes, J., Rodin, J., Rangan, V.K., Hesselbein, F. (2008). *The Five Most Important Questions You Will Ever Ask About Your Organization*. Jossey-Bass, San Francisco.

ENCODE Project Consortium (2004). The ENCODE (ENCyclopedia of DNA Elements) Project. *Science*, 306(5696), 636–640.

ENCODE Project Consortium (2007). Identification and analysis of functional elements in 1% of the human genome by the ENCODE pilot project. *Nature*, 447(7146), 799–816.

European Commission DR (2010). Stratification biomarkers in personalised medicine. Workshop report, European Commission.

Gerstein, M.B., Bruce, C., Rozowsky, J.S., Zheng, D., Du, J., Korbel, J.O., Emanuelsson, O. et al. (2007). What is a gene, post-ENCODE? History and updated definition. *Genome Research*, 17(6), 669–681.

Gilbert, W. (1978). Why genes in pieces? *Nature*, 271(5645), 501.

Good, I.J. (1965). Speculations concerning the first ultraintelligent machine. *Advances in Computers*, 6, 31–88.

Goodman, L.L., Wintrobe, M.M., Dameshek, W., Goodman, M.J., Gilman, A., McLennan, M.T. (1946). Nitrogen mustard therapy: Use of methyl-bis(beta-chloroethyl)amine hydrochloride and tris (beta-chloroethyl)amine hydrochloride for Hodgkin's disease, lymphosarcoma, leukemia and certain allied and miscellaneous disorders. *JAMA*, 132, 126–132.

Harrington, L.E., Hatton, R.D., Mangan, P.R., Turner, H., Murphy, T.L., Murphy, K.M., Weaver, C.T. (2005). Interleukin 17-producing CD4+ effector T cells develop via a lineage distinct from the T helper type 1 and 2 lineages. *Nature Immunology*, 6, 1123–1132.

van Helden, P. (2013). Data-driven hypotheses. *EMBO Rep.*, 14(2),104.

Heudel, P., Livartowski, A., Arveux, P., Willm, E., Jamain, C. (2016). ConSoRe : un outil permettant de rentrer dans le monde du *Big Data* en santé. *Bull. Cancer*, 103, 949–950.

Hey, T., Tansley, S., Tolle, T. (2009). *The Fourth Paradigm: Data-Intensive Scientific Discovery*. Microsoft Research/Redmond, Washington.

ICGC, T., Hudson, J., Anderson, W., Artez, A., Barker, A.D., Bell, C., Bernabé, R.R. et al. (2010). International network of cancer genome projects. *Nature*, 464(7291), 993–998.

International HapMap Consortium, Frazer, K.A., Ballinger, D.G., Cox, D.R., Hinds, D.A., Stuve, L.L., Gibbs, R.A. et al. (2007). A second generation human haplotype map of over 3.1 million SNPs. *Nature*, 449(7164), 851–861.

International Human Genome Sequencing Consortium (2004). Finishing the euchromatic sequence of the human genome. *Nature*, 431(7011), 931–945.

Itik, M. and Banks, S.P. (2010). Chaos in a three dimensional cancer model. *IJBC*, 20, 71–79.

Lander, E.S., Linton, L.M., Birren, B., Nusbaum, C., Zody, M.C., Baldwin, J., Devon, K. et al. (2001). Initial sequencing and analysis of the human genome. *Nature*, 409(6822), 860–921.

Lapham, E.V., Kozma, C., Weiss, J.O., Benkendorf, J.L., Wilson, M.A. (2000). The gap between practice and genetics education of health professionals: HuGEM survey results. *Genetics in Medicine*, 2(4), 226–231.

Laugier, A. (1996). Le premier siècle de la radiothérapie en France. *Bulletin de l'Académie nationale de médecine*, 180(1), 143–60.

Le Tourneau, C., Delord, J.P., Gonçalves, A., Gavoille, C., Dubot, C., Isambert, N., Campone, M. et al. (2015). Molecularly targeted therapy based on tumour molecular profiling versus conventional therapy for advanced cancer (SHIVA): A multicentre, open-label, proof-of-concept, randomised, controlled phase 2 trial. *Lancet Oncology*, 16(13), 1324–1334.

Leary, R.J., Kinde, I., Diehl, F., Schmidt, K., Clouser, C., Duncan, C., Antipova, A. et al. (2010). Development of personalized tumor biomarkers using massively parallel sequencing. *Science Translational Medicine*, 2(20).

LeCun, L., Bengio, Y., Hinton, G. (2015). Deep learning. *Nature*, 521(7553), 436–444.

Lee, B.T., Barber, G.P., Benet-Pagès, A., Casper, J., Clawson, H., Diekhans, M., Fischer, C. et al. (2022). The UCSC genome browser database: 2022 update. *Nucleic Acids Research*, 50(D1), D1115–D1122.

Letellier, C., Denis, F., Aguirre, L.A. (2013). What can be learned from a chaotic cancer model? *Journal of Theoretical Biology*, 322, 7–16.

Levy, S., Sutton, G., Ng, P.C., Feuk, L., Halpern, A.L., Walenz, B.P., Axelrod, N. et al. (2007). The diploid genome sequence of an individual human. *PLoS Biol.*, 5(10).

Ley, T.J., Mardis, E.R., Ding, L., Fulton, B., McLellan, M.D., Chen, K., Dooling, D. et al. (2008). DNA sequencing of a cytogenetically normal acute myeloid leukaemia genome. *Nature*, 456(7218), 66–72.

Lieverse, A.R., Temple, D.H., Bazaliiskii, V.I. (2014). Paleopathological description and diagnosis of metastatic carcinoma in an Early Bronze Age (4588+34 Cal. BP) forager from the Cis-Baikal region of eastern Siberia. *PLoS ONE*, 9(12).

Lupski, J.R., Reid, J.G., Gonzaga-Jauregui, C., Rio Deiros, D., Chen, D.C., Nazareth, L., Bainbridge, M. et al. (2010). Whole-genome sequencing in a patient with Charcot-Marie-Tooth neuropathy. *The New England Journal of Medicine*, 362, 1181–1191.

Ma, P.C., Zhang, X., Wang, Z.J. (2006). High-throughput mutational analysis of the human cancer genome. *Pharmacogenomics*, 7(4), 597–612.

Manolio, T.A., Collins, F.S., Cox, N.J., Goldstein, D.B., Hindorff, L.A., Hunter, D.J., McCarthy, M.I. et al. (2009). Finding the missing heritability of complex diseases. *Nature*, 461(7265), 747–753.

Mardis, E.R. (2009). New strategies and emerging technologies for massively parallel sequencing: Applications in medical research. *Genome Medicine*, 1(4), 40.

Mardis, E.R. and Wilson, R.K. (2009). Cancer genome sequencing: A review. *Human Molecular Genetics*, 18, R163–R168.

Mardis, E.R., Ding, L., Dooling, D.J., Larson, D.E., McLellan, M.D., Chen, K., Koboldt, D.C. et al. (2009). Recurring mutations found by sequencing an acute myeloid leukemia genome. *The New England Journal of Medicine*, 361(11), 1058–1066.

Martincorena, I., Raine, K.M., Gerstung, M., Dawson, K.J., Haase, K., Van Loo, P., Davies, H. et al. (2017). Universal patterns of selection in cancer and somatic tissues. *Cell*, 171(5), 1029–1041.

Mateo, J., Carreira, S., Sandhu, S., Miranda, S., Mossop, H., Perez-Lopez, R., Nava Rodrigues, D. et al. (2015). DNA-repair defects and olaparib in metastatic prostate cancer. *The New England Journal of Medicine*, 373(18), 1697–1708.

Mattick, J.S. (2010). The central role of RNA in the genetic programming of complex organisms. *Anais da Academia Brasileira de Ciencias*, 82(4), 933–939.

McCulloch, W. and Pitts, W. (1943). A logical calculus of the ideas immanent in nervous activity. *Bulletin of Mathematical Biophysics*, 5, 115–133.

Murrell, A., Rakyan, V.K., Beck, S. (2005). From genome to epigenome. *Human Molecular Genetics*, 14, R3–R10.

Navin, N., Kendall, J., Troge, J., Andrews, P., Rodgers, L., McIndoo, J., Cook, K. et al. (2011). Tumour evolution inferred by single-cell sequencing. *Nature*, 472, 90–94.

Nelson S.B, Sugino, K., Hempel, C.M. (2006). The problem of neuronal cell types: A physiological genomics approach. *Trends in Neurosciences*, 29(6), 339–345.

Nielsen, C.B., Cantor, M., Dubchak, I., Gordon, D., Wang, T. (2010). Visualizing genomes: Techniques and challenges. *Nature Methods*, 7(3), S5–S15.

Ormond, K.E., Wheeler, M.T., Hudgins, L., Klein, T.E., Butte, A.J., Altman, R.B., Ashley, E.A., Greely, H.T. (2010). Challenges in the clinical application of whole-genome sequencing. *Lancet*, 375(9727), 1749–1751.

Pleasance, E.D., Cheetham, R.K., Stephens, P.J., McBride, D.J., Humphray, S.J., Greenman, C.D., Varela, I. et al. (2010a). A comprehensive catalogue of somatic mutations from a human cancer genome. *Nature*, 463(7278), 191–196.

Pleasance, E.D., Stephens, P.J., O'Meara, S., McBride, D.J., Meynert, A., Jones, D., Lin, M.L. et al. (2010b). A small-cell lung cancer genome with complex signatures of tobacco exposure. *Nature*, 463(7278), 184–190.

Pushkarev, D., Neff, N.F., Quake, S.R. (2009). Single-molecule sequencing of an individual human genome. *Nature Biotechnology*, 27(9), 847–852.

Rosenblatt, F. (1958). The perceptron: A probabilistic model for information storage and organization in the brain. *Psychological Review*, 65(6), 386–408.

Schmidhuber, J. (2015). Deep learning in neural networks: An overview. *Neural Networks*, 61, 85–117.

Schneider, V.A., Graves-Lindsay, T., Howe, K. (2017). Evaluation of GRCh38 and de novo haploid genome assemblies demonstrates the enduring quality of the reference assembly. *Genome Research*, 27(5), 849–864.

Schuster, S.C. (2008). Next-generation sequencing transforms today's biology. *Nature Methods*, 5(1), 16–18.

Sethi, I. (1990). Entropy nets: From decision trees to neural networks. *Proceedings of the IEEE*, 78(10), 1605–1613.

Smith, A., Carter, E.B., Herrick, A.D. (1948). *Drug Research and Development.* Revere Pub. Co., New York.

Sontag, S. (1990). *Illness as Metaphor and AIDS and Its Metaphors.* Penguin, London.

Stein, L.D. (2010). The case for cloud computing in genome informatics. *Genome Biology*, 11(5), 207.

Tang, F., Barbacioru, C., Wang, Y., Nordman, E., Lee, C., Xu, N., Wang, X. et al. (2009). mRNA-Seq whole-transcriptome analysis of a single cell. *Nature Methods*, 6, 377–382.

The 1000 Genomes Project Consortium (2010). A map of human genome variation from population-scale sequencing. *Nature*, 467(7319), 1061–1073.

The 1000 Genomes Project Consortium (2015). A global reference for human genetic variation. *Nature*, 526(7571), 68–74.

Valsecchi, M.E., Díaz-Cantón, E., de la Vega, M., Littman, S.J. (2014). Recent treatment advances and novel therapies in pancreas cancer: A review. *Journal of Gastrointestinal Cancer*, 45(2), 190–201.

Vaquerizas, J.M., Kummerfeld, S.K., Teichmann, S.A., Luscombe, N.M. (2009). A census of human transcription factors: Function, expression and evolution. *Nature Reviews Genetics*, 10(4), 252–263.

Venter, J.C., Adams, M.D., Myers, E.W., Li, P.W., Mural, R.J., Sutton, G.G., Smith, H.O. et al. (2001). The sequence of the human genome. *Science*, 291(5507), 1304–1351.

Walter, M.J., Graubert, T.A., Dipersio, J.F., Mardis, E.R., Wilson, R.K., Ley, T.J. (2009a). Next-generation sequencing of cancer genomes: Back to the future. *Personalized Medicine*, 6(6), 653.

Walter, M.J., Payton, J.E., Ries, R.E., Shannon, W.H., Deshmukh, H., Zhao, Y., Baty, J. et al. (2009b). Acquired copy number alterations in adult acute myeloid leukemia genomes. *PNAS*, 106(31), 12950–12955.

Complexity or Complexities of Information: The Dimensions of Complexity

If we do not organize complexity, complexity will destroy us, slowly but surely.

Jacques PRINTZ

5.1. Introduction

The sciences of complexity related to information processing were born in the context of the first large computerized systems of the 1950s, and more particularly, from the development of the first computers. With the appearance on the technological scene of these strange, completely new objects, programs – that is to say, algorithms – and the individuals responsible for their creation, that is to say, programmers, the study of complexity has become a necessity, if not a question of economic survival.

For the systems architect, the main question is: how can we organize this complexity, or even make it an ally, so that the system created remains under the control of its designers and operators? In short, engineering should be employed as it has always been: at the service of users who must have confidence in what they handle and be aware of the risks they really incur by using it, which requires a greater or shorter learning process.

Since then, the problem of complexity has spread almost everywhere, in the spheres of biology and in the human sciences, such as economics and/or business management, each with its own specificities. It is therefore all the more important to

Chapter written by Jacques PRINTZ.

know the context in which it was born and the lessons that can be drawn from its initial uses before going further (for greater detail, see Printz 2012, 2018).

5.2. A brief historical overview

The word "complexity" itself does not feature in Lalande's philosophical dictionary. It does appear in the *Dictionnaire des sciences* (Flammarion, 1997), edited by Michel Serres and Nayla Farouki. However, it is absent from the *Vocabulaire technique et analytique de l'épistémologie* (PUF, 1999, by Robert Nadeau, a professor at the University of Quebec).

The article on "Complexité" on the French Wikipedia has two pages of useful information but does not mention complexity as such from the point of view of computer systems engineering; the English version, "Complexity", is a little more extensive, with six useful pages with references; in particular, a reference to the Santa Fe Institute[1], a leading research institution in this new field which is not mentioned in the French article.

If one searches for the words "complex/complexity" on France's favorite search engine, one immediately comes across Edgar Morin and Jean-Louis Le Moigne, who saturate the search space, and everything revolves around the six volumes of *La Méthode*, published by Edgar Morin from 1977 onwards. In summary, the results are meager and require a real effort in terms of reading and research on the documentation to go back to the sources: von Neumann, Shannon, Turing and Wiener.

What we now call "complexity" entered the scientific field through the work of Maxwell and Boltzmann on the kinetic theory of gases, whose ambitious goal was to describe the macroscopic behavior of gases (pressure, volume, temperature, entropy, etc.) from the atomic particles that constitute them, which are very large in number, following on from the work of Avogadro – the Avogadro constant NA^2, a number of 23 digits, that is, one hundred thousand billion billions, which in the usual notation is 6.02×10^{23} water molecules held in 18 g of water. However, neither Boltzmann, Maxwell, nor anyone else mentions complexity.

To explain the homogeneity of a gas mixture, Boltzmann had to calculate the number of possible configurations that a gas, confined in an enclosure, could take,

---

1 See https://en.wikipedia.org/wiki:Santa_Fe_Institute.

2 See https://en.wikipedia.org/wiki/Avogadro_constant.

numbers that he said "defied the imagination" – these numbers were manipulated by the little intelligent being created by Maxwell, since then called "Maxwell's demon" because it defies the second principle of thermodynamics – however, they are numbers that will be found in Boltzmann's famous formula $S = k \times \log (W)$, which links the entropy of a mixture with the number of configurations that this mixture can take (the number W is used for the German *Wahrscheinlichkeit*, used for the calculation of probabilities; mathematicians use the Greek letter Ω to describe the same thing). k is Boltzmann's constant, a very small number, while W, expressed in logarithmic units, is itself very large according to one of the formulas of infinitesimal calculus, where $0 \times \infty \rightarrow n$, n being a finite number – the infinitely small and the infinitely large neutralize one another – under certain conditions that the calculus specifies.

The mathematician Émile Borel, a great specialist in probabilities, who made a sensational entry into the very heart of the most reductionist deterministic physics, published several studies on these numbers, which he described as "imagined" or "inaccessible". He used them in a metaphor known as the "dactylographic monkeys"– see his book, *Le Hasard* (1911) – to explain what is described as "impossible" in the calculation of probabilities, that is, a very low probability, below a threshold that he would specify.

In the 1920s and 1930s, the evolutionary biologists of this period – at least some of them – used this metaphor to show that chance alone could not account for the creation of proteins, which are highly organized assemblies of atoms, and a fortiori, according to them, the living beings and lineages that paleontologists were beginning to identify, which were a gigantic puzzle. It is in this context that Teilhard de Chardin, himself a geologist and paleontologist, spoke of the "infinitely complex", defined as a third infinity, besides the infinitely large one of astrophysics and the infinitely small one encountered in the atomic world, which were revealed in this era.

However, all this remains terribly qualitative, if we put aside the work of Borel, who also does not speak of complexity. With computers, von Neumann at the helm, and the development of the first large technological systems, this starts to change radically in the 1940s and 1950s.

5.3. The phenomenology of complexity in systems engineering

This section complements the generalizations that have been presented in our book (Printz 2020). Architects of computerized systems created by human intelligence face two main types of complexity-related challenges:

– A static complexity that has long been understood through the mechanisms of modularity and hierarchization; that is, the constituent parts of the system, still called "elements", which have been at the heart of information and communication technologies (ICT) since the beginning. Hierarchizing, classifying, researching, creating/deleting and restructuring are the basic operations at the heart of ICT and the ordering of our knowledge (often summarized, among computer scientists, by the acronym CRUDE (create, retrieve, update, delete, execute) for the five fundamental generic operations/instructions). This is the ontological aspect of complexity, which we will call "complication", as in watches of the same name or in robotic assembly lines in the automotive industries (i.e., a "simple" nomenclature of parts used to generate traceability), avionics or electronic chips. In systems engineering, this is what is called "configuration management", for which there are effective support systems and management tools.

– A dynamic, behavioral complexity, more subtle because it results from a combinatorial of memory states and/or past situations influencing the present, which represent immense inaccessible numbers (see below) when not controlled by the architecture.

This complexity is related to:

– the evolution of the organizational and human environments, the users and the engineering teams (the three modes of existence: an abbreviation for users, system and engineering; see Figure 5.1), which requires more and more interactivity;

– the evolution of the system itself and of its construction process – the assembly plan – which become more precise and adaptable during the realization (referred to as "agility" within the systems engineering community);

– the evolution of the technology, which today allows for massive asynchronism in the transformations carried out by the system and task parallelism.

The "initial lifeline" of the first systems gives rise to hundreds, and now thousands, of lifelines – called threads in the jargon – that synchronize in order to perform the service requested by users (see Figure 5.1).

This is a complexity that materializes:

– from the flow of transactions (these are "atomic" operations) that we can observe, amounting today to tens of thousands per second, in real time moreover, transactions programmed with the five generic CRUDE operations – that is, those of a Turing machine (TM);

– from the increasingly high integration costs (i.e., an "effort" expressed in working hours, provided by the human environment) of the platforms, which are the true indicator of this complexity.

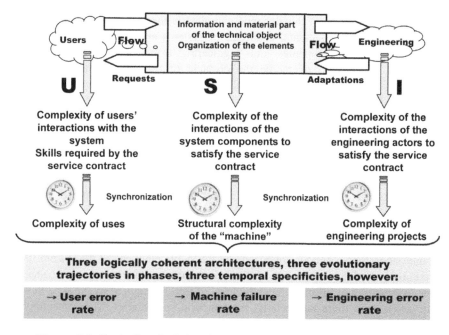

Figure 5.1. *Illustration depicting the three areas of complexity. For a color version of this figure, see www.iste.co.uk/briffaut/complexities1.zip*

Big data and cloud computing, implemented in data centers, are the consumer marketing image of this irreversible evolution, but even more so, the massive parallelism that is manifesting itself in the latest generations of multi-core chips and machines specializing in intensive computing, such as Bull-ATOS's Tera-100/1000 machines, built with eight-core chips, while chips with several hundred cores are being studied and/or already being experimented with; this will upset best practices that have been used for sequential programming since the beginning.

The diagram in Figure 5.1 summarizes the problematic of the three modes of existence; it highlights the different actors of complexity, which makes it possible to understand the phenomenology in its globality, both subject and object in their respective roles, which can be exchanged. The logic of analysis used here follows the founding work on information by Gilbert Simondon[3]. The purpose of the system, its service contract, which is indispensable to its stability, is carried out according to the needs and requirements of its users and stakeholders.

---

3 See, in particular, his thesis *Du mode d'existence des objets techniques* (Aubier, re-issue, 2012); and *L'Individuation à la lumière des notions de forme et d'information* (Millon, re-issue, 2013).

It is the organization of this dynamic complexity, for ever more secure systems, which is today the fundamental challenge of systems engineering, the structuring divide between the companies and engineering teams that will develop the architectural capacity to overcome the obstacle of this complexity by training themselves seriously and those who will stumble through ignorance or technical "illiteracy". The watchword proclaimed by all the actors in engineering is *agility*, but the difficulty of "agile" engineering is obviously in *how* to implement it practically, organizing it through a reasoned and rigorous architecture[4] of the modules and their internal and external interfaces[5].

To make the evolutionary trajectories logically coherent and secure, and to improve the service contract with users, is to realize negotiated couplings between all the actors of these trajectories, through the contractualized interfaces, which, as we will see, increases complication, but strongly decreases complexity, the terms being those of the CCUI concept (complication, complexity, uncertainty, ignorance). The interfaces, hierarchically organized, are the stable elements of the system (they are the axioms of the system, its semantic invariants), which permit the exchange and transformation of information at the level of finesse required by the specificities of the actors and, more particularly, the temporal specificities. More interactivity and agility entail finer division of functions (hence indispensable modularity), necessitate more interacting elements (hence finesse of regulations and controls, reliability/robustness/resilience, security), impose more difficult integration for production platforms when considering political, economic, social, technological, environmental and legal (PESTEL[6]) factors and, therefore, lead to more expensive integration. Here again, the puzzle metaphor remains one of the most relevant to understanding the nature of the problem for the organization of complexity.

5.3.1. *Measuring the complexity of an assembly through the integration process and tests*

This is the IVVT activity in the systems engineering process terminology[7].

---

4 See Sifakis, J. (2012). *Rigorous System Design*. EPFL, Lausanne. See also our publication *Architecture logicielle* (Printz 2012).

5 See Krob, D. (2014). *Complexité-Simplexité. Lectures at the Collège de France*. Available at: http://books.openedition.org/cdf/3388?lang=fr.

6 See https://en.wikipedia.org/wiki/PEST_analysis; as a complement, see our work *Estimation des projets de l'entreprise numérique*, which explores this in detail.

7 See the white paper *Intégration et Complexité*. Available at: http://cesam.community/wp-content/uploads/2017/07/121017-LB-IC-Synthèse.pdf.

The *Integration and Complexity* research seminar, held in 2012 (with the support of the CESAMES association and the "Complex Systems" chair at the École polytechnique), had the objective of clearly stating the problem of integration, that is, the construction process, which is the fundamental foundation on which the digital company, and a fortiori the digital society, will be able to develop at the right speed while controlling the risks, the potentialities, that technological innovation (production platforms, processing capacities, "connected" objects, etc.) places at its fingertips to satisfy emerging user needs.

The primary objective of the seminar was to validate the relevance of a measure of complexity based on integration tests, that is, a set of texts proving in an acceptable, if not irrefutable, manner that the system taken as a whole satisfies the requirements of the service contract of the end-users of the system, including dynamic behaviors and PESTEL factors. We see that this is indeed the case insofar as these tests are directly derived from the need and the design, which determines the assembly plan, expressed with an ad hoc notation[8], and the parameters implemented to build the system.

Let us recall that, for a watch with complications, for example, the Patek Philippe caliber 89 with 1,728 parts, the number of combinations in the assembly is the factorial of 1,728, that is, $1,728 = \pm 1.07 \times 10^{4,646}$, obtained via Stirling's formula; that is, a number so large that it cannot be written down, even if we use all the hydrogen atoms in the visible universe, that is, 10^{80}. In order to exhaust the trial-and-error combinatorics of such a number, assuming that one is able to make one trial per Planck time unit, that is, 10^{-43} seconds, the age of the universe estimated at 13.7 billion years (about 10^{17} s and consequently 10^{60} Planck units) would not suffice, and not by a long shot! There is thus an obvious problem with "randomness" in the general sense of the term.

It is also the method used to test very-large-scale integration (VLSI)/self-organized criticality (SOC) in accordance with the hardware engineers' principle of design-to-test architecture. In VLSI, the fundamental term is I, for integration, which in the case of SOCs is a process with several hundred steps, resulting in SOC build times in the region of 6–9 months.

The word "complexity"[9] is endowed with a rich polysemy that can make its use dubious if one does not take some terminological precautions. In the world of computerized systems, it can be defined by making a common sense distinction between the number of elements (modules) that make up the system (complication,

---

8 Activity diagrams, the BPMN/BPEL language, etc.
9 See the paper by Nancy Leveson of MIT, "Complexity and Safety", presented at CSDM2011.

i.e., a simple count, such as the number of *transistors* integrated in a VLSI, the number of instructions in a program, etc.) and the number of static and dynamic relationships that these elements have with each other, and with the outside world, while the system "lives" (the "real" complexity, i.e., the connectivity, the explicit or implicit functional dependencies, etc.).

For biologists since Ramon y Cajal[10] (1852–1934) or the neurologist Paul Broca[11] (1824–1880), the complexity of a living organism has been evaluated according to the mass of the organism's brain in relation to its size, referred to as the "degree of cephalization", as explained in Figure 12 of Changeux's book, *L'Homme neuronal* (1983). This places man at the top of the hierarchy, with a brain that has about 100 billion neurons – a figure that has varied a lot since the 1970s and 1980s, when it was estimated at 10–15 billion – and 10^{15} connections between neurons. Since the development of methods for sequencing the DNA molecules that make up the chromosomes of all living organisms, biologists have been using the size of genomes[12] more frequently. The human genome is estimated at 3.4 billion letters, or base pairs (ATCG), but it is not the largest known genome. Knowing that a 400-page book is about 1 million characters, our genome in this format, contained within each of our cells, is equivalent to a library of 3,400 books.

In order to apprehend all the facets of complexity, taking care to not separate them into two aspects of the same reality, we must first distinguish the complication aspect, that is, its simple enumeration, and the relations/couplings aspect, which develops over time. We will reserve the use of the words "complexity" and "complex" for the enumeration aspect of the relations/couplings, which takes into account the dynamic dimension of the interactions between modules, which will be materialized by *integration tests*, at one moment or another within the lifetime of the system.

From the white paper resulting from the *Integration and Complexity* seminar, and in our previous work *Architecturelogicielle et Estimation des projets de l'entreprise numérique* (Software Architecture and Estimation of Digital Enterprise Projects), we can say in summary that a system is:

– Complex if there are many relationships between the elements that make up the system and the environment within which it is integrated. Counting the relationships/couplings (in systems engineering, this is done with N2 matrices) is a more difficult exercise than counting the elements, since there are clearly visible,

10 See https://en.wikipedia.org/wiki/Santiago_Ram%C3%B3n_y_Cajal.

11 See https://en.wikipedia.org/wiki/Paul_Broca.

12 See, for example, Smith, J.M. and Szathmary, E. (1995). *The Major Transitions in Evolution*. Oxford University Press, Oxford; and more recently: Dessalles, J.-L. (2016). *Le Fil de la vie*. Odile Jacob, Paris.

explicit relationships and hidden, implicit relationships, which will only be revealed with use: this is the phenomenon known as "emergence", which is well-known in the theories of complexity (dynamic systems) and condensed matter physics. The failure of a system is a dreaded emergent phenomenon that results from an inconsistency in the global state of the system, which was not visited during validation, verification and testing, that is, during integration. A frequent case of this type of situation appears with the "wild" mutualization of functions and interoperability in the problematic of system of systems.

– Complicated if it includes many elements. Hence, the importance of typology and simply volumetric analyses (i.e. the nomenclatures of the parts and/or components of a piece of equipment): we count the instructions of the programs, the number of calculation steps to solve a problem (in theoretical computer science, this is the "algorithmic complexity"), the defects and/or the adaptations (anomaly reports (ARs)) observed and the resulting corrective actions (CAs: periodic updates, patches, etc.), the computer equipment, the programmers, the error messages and online help, the users (for the "scalability" of the system, that is, its capacity to grow without degrading performance), etc. Nevertheless, we often forget to count the possible states of the system (evolution of its configuration), which are however the preponderant factor of complexity, that is, the memory states resulting from the actions carried out in the past which influence the present (feedback and/or hysteresis).

The most recent complexity theories have highlighted the importance of textual complexity (or descriptive complexity, in the form of the graphical diagrams used for a long time in engineering, such as the IDEF language inspired by the first diagrams used by Forrester, which can always be reduced to Kolmogorov Chaitin (KC) text[13,14]), associated with a particular machine, to measure the quantity of information. From a theoretical point of view, we can be satisfied with TMs, which makes the TM and its generic CRUDE operations the "natural" standard of complexity (a terminology used by the mathematician-logicist Jean-Yves Girard)[15], that is, its *scale*.

KC textual complexity is the length of the shortest program that satisfies all and only the requirements defining a system, notwithstanding its organization, which has an energy cost to be minimized. It is a fundamental measure.

---

13 The language of our printers, PDF, makes it possible to print both text and/or drawings. See https://en.wikipedia.org/wiki/Portable_Document_Format.

14 See the two volumes *L'Héritage de Kolmogorov enphysique* and *L'Héritage de Kolmogorov en mathématiques*; his main works are translated into French and commented on by Belin (2003 and 2004).

15 See Turing, A. and Girard, J.-Y. (1999). *La Machine de Turing*. Points Sciences; in particular, Girard's commentary on the famous Turing article.

A complicated system has a low KC complexity. A large system like the Great Wall of China will have a low KC; its pattern is re-patterned. The effort/energy relation needed to produce such a system is *linearly related to the size.*

In the general case, the relation is a power law, or even an exponential. Fermat's great theorem has a low KC complexity, but if we take into account the length of the proof elaborated by Andrew Willes – several hundreds of very dense pages – we land back on our feet!

In the real world of systems, the TM is replaced by a real machine, in the so-called von Neumann architecture, that is, an abstracted machine, either universal, with languages adapted to the needs of programmers, or specialized with ad hoc languages (the domain specific language (DSL)) such as exist for databases (the SQL language), networks (the language of the http interface of the Web with its eight commands) and other devices; in general, any API library that satisfies a well-identified business need. For von Neumann, one could grasp the entirety of mathematical knowledge with a dozen or so symbols found in programming languages, a figure that was considered low and which triggered hilarity among the participants of a conference he addressed for the ACM, to which he replied: "If people do not believe that mathematics is simple, it is only because they do not realize how complicated life is". *Complicated*, he said!

KC textual complexity thus emphasizes the importance of the underlying machine (see the case of DSLs with their specialized machines) as a yardstick for quantifying information content; in fact, this definition of KC complexity is the foundation of cost estimation models used in engineering. Textual complexity is thus a relative notion related to the scale of the calibration, that is, to the expressive power of the underlying low-level language (LLL) machine.

5.4. The four dimensions of complexity

We will therefore distinguish four notions, or aspects, correlated in complexity (we could also say intricated, to use a term from quantum mechanics to express duality/complementarity, except that here, there are four of them), namely, complex, complicated (as recalled above), uncertainty and ignorance[16], often associated with ordinary complexity to characterize the approximate side of any modeling; thus, in

---

16 See the notion of probably approximately correct (PAC) developed by Valiant in his book by the same name. In his approach, he distinguishes phenomena for which there is a theory, and therefore associated models, from those for which there is no theory, just data, possibly in large masses; hence, the theory-full/theory-less duality. See also the book by David Ruelle, *Hasard et chaos*, one of the pioneers in studies on chaos.

the end, a composite "vector" complexity in order to take into account the four dimensions with which we are going to analyze the external environment of the systems via the PESTEL factors.

The uncertainty of the behaviors and the ignorance, where we have certain possible configurations, are directly linked to the combinatoriality induced by the space of the possible states of the system, taking into account the hazards of the environment and the organic structure of the system, including the failures which can occur. It is important to realize that this combinatorial system quickly defies imagination, because there are N interacting systems, each of which can be in s configurations (e.g. nominal, faulty, uncertain, depending on the modalities), that is to say an exponential of order s^N, in decimal notation $10^{N \times \log(s)}$.

If the system memorizes its states and these memorized states are used to make future choices, then the combinatorics of possible configurations will grow exponentially.

In the first step, we must consider the set of parts that can be formed by N interacting elements, thus, 2^N; in the second step, these interactions will be reinjected into the set of new parts to be considered, that is, a new combinatorial of 2^{2N}, and so on; thus, after a certain number of iterations, $2**2**...**2**n$, which is a function whose growth is ultra-rapid, a bit like the Ackermann function, well-known to computer scientists. This is the reason why we have to be very careful with memorization, and why microprocessors use $stateless$[17] circuits, to guarantee testability. This is also the reason why, on the software and program side, it is imperative to minimize and monitor the memories shared between the different processes. In the event of an error, it must be possible to retrace the path that led to the failure. This is both a considerable and inexhaustible source of complexity.

For example, a shared memory of 1 megabyte, or 1,048,576 bytes, can take a number of configurations equal to $2,561,048^{576}$, which is equivalent to $10^{\log s \ (256) \times 1,048,576}$, that is, $10^{0.63} \times 10^{2.525.222}$, or $4.27 \times 10^{2.525.222}$, which is still an inaccessible number[18].

Hence, the need to organize these interactions with hierarchical structures, which are easy to implement, as suggested by Simon in his seminal work, already cited here, *Sciences of the Artificial*, in Chapter 8 concerning "The architecture of complexity".

---

17 See Deming, P. and Lewis, T. (2017). Exponential laws of computing growth. *Communications of the ACM*, 60(1); this article insists once again on this fundamental point.
18 See https://en.wikipedia.org/wiki/%C3%89mile_Borel.

In addition to the complex/complication aspects, so as to take into account the theoretical or empirical knowledge that we can have of the phenomena that we wish to use to build the systems, we must add two new aspects: uncertainty and ignorance.

– *Uncertainty*: the expected behaviors are generally uncertain, unseen, counterintuitive; for example, performance, load control, safety, robustness. However, we can experiment via ad hoc or in situ models and observe with probes to detect breakpoints (equivalent to symmetry "breaks" that accompany phase changes in physical systems). We then have a statistical "truth" of the expected uncertain behaviors. In complexity/emergence theories, it is the phenomenon of SOC (introduced by the physicist Per Bak), that automatically produces a complexity of this type, but one where a stable statistical order emerges, such as deterministic chaos (this aspect corresponds to Valiant's notion of *theory-full*). An excellent example of a simple law that can generate chaotic phenomena is in the article by the ecological biologist Robert May, "Simple mathematical models with very complicated dynamics" (1976), available online.

– *Ignorance* or lack of knowledge: the system can be the seat of emerging phenomena as yet unknown, favorable[19] and/or unfavorable to the service contract, generally coming from the hardware part of the system (wear and tear, aging, etc.), but also coming from the software if the interfaces are poorly controlled and permit the propagation of inconsistent memory states and hardware/software interactions generating unverifiable combinatorics. For example, we do not know how very-high-integration chips (8–10 billion transistors for systems on chips) will age, and we do not know anything about the nature of the failures that this aging will cause: their use is currently forbidden in equipment with a lifetime >2–3 years. At best, we can introduce a nominal behavior model (i.e. an invariant) that will be monitored and tested periodically, the test itself having to be safe, in order to observe and understand what is happening. Recall that memory of size N shared among several processes, combined with the processing capabilities of massive parallelism, mechanically manufactures a colossal combinatorics, of the order $8N$, or in decimal notation $10^{N \times log(8)} \approx 10^{0.9 \times N}$; this is such an immense number that it defies imagination, given that some systems integrate thousands of interacting modules. This also opens up the possibility of new applications that one has no idea of, given the new capabilities thus offered (this aspect corresponds to Valiant's *theory-less* concept). In engineering, examples of this type are innumerable, hence the integrated monitoring systems that are an integral part of the design, also known as

---

19 The physical phenomenon of semi-conductivity is typical of the emergence of radically new and useful properties, as it occurs only in single crystals of silicon whose lattice is very slightly disturbed by carefully measured impurities.

system management or autonomic computing (used by IBM), that is, a system that supervises another system.

The presence of non-reproducible residual errors, therefore inevitable, equivalent to "noise" in the theory of information, in large-scale software mechanically generates a combination of uncertainties/ignorances that will lead to more or less serious failures; in the jargon of computer scientists, Jim Gray[20] distinguished what he called the *Bohr-bugs*, in reference to the Newtonian model of the Bohr atom, that is to say the "good" errors, simple to diagnose, and the *Heisen-bugs*, the most vicious errors because they are random, like the quantum phenomena analyzed by W. Heisenberg, and therefore not reproducible. By investing 70–80% of the development effort in IVVT, NASA has reached rates of 0.1 error per thousand lines of code, which means that for a software of 1 million lines, there are about 100 "stray" errors. We know that there are still errors, their number is an increasing function of the textual complexity KC, but we know neither where they are nor what the consequences of their manifestation as a failure in such and such a situation will be (see the phenomena of SOC as described by the physicist Per Bak, already cited above). In computerized systems, the notion of autonomic computing (i.e. of self-organized modules that depend only on themselves, by construction) becomes vital from a certain critical size (controllability), as well as the service processors for monitoring and repair (recovery). It is a component of dependability that plays a crucial role in the dimensioning of systems.

Rigorous methods, such as model checking, recently honored by a Turing Award[21], allow us to ensure that certain feared events, that is, a particular possible property that can be modeled, are unattainable through construction (which we can verify/validate in the system integration phase). Nevertheless, we will never have absolute certainty about the globality of the system. However, for a particular system or a class of systems, we can, through construction, by organizing the system and its modules hierarchically, control the combinatorics and maintain a testability invariant at all levels of the construction, that is, a testable architecture, or design a test of the complex circuit integrators (VLSI/SoC) already mentioned, which means that we know exactly what we are doing and that fault diagnosis is always feasible, which must remain a basic requirement of any system that is the fruit of human intelligence and ingenuity. Highly complex systems, such as our smartphones, are an existential proof of this capability.

---

20 See https://en.wikipedia.org/wiki/Jim_Gray_(computer_scientist); Turing Award, 1998.
21 In 2007 to Clarke, E., Emerson, E. and Sifakis, J. See http://amturing.acm.org/byyear.cfm.

5.5. The term "simplexity": A remark on Richard Feynman's Nobel lecture

In his 1966 Nobel Lecture, Richard Feynman made an interesting remark that illustrates his pedagogical genius as well as his creativity, at the periphery of popular trends and somewhat off the beaten track. He states quite simply:

> It always seems strange to me that the laws of physics, when discovered, come in so many different forms which do not at first appear to be equivalent, but which can then be shown to be so by bringing mathematics into play. […] Perhaps a thing is simple if you can describe it fully in several different ways without immediately knowing that you are describing the same thing.

He tells us that this is perhaps a way of defining simplicity, which is another way of defining complexity, by taking the converse, that is to say, "That which is complex is that which is not simple. That which is simple is what can be described in different ways".

In addition to computer science, which offers, by means of multiple languages and/or equivalent representations that it routinely uses, countless examples of this diversity of formulations of the same algorithmic transformation, mathematics itself, and more particularly geometry, can help illustrate the depth of Feynman's intuition, who also said, "What I cannot create, I do not understand"[22]. This constructive way of defining complexity is consistent with KC textual complexity, since a "good" language must reduce textual complexity.

On these different points of view, it is relevant to refer to the fundamental distinction between intension versus extension found in set theory or to that of external/internal languages introduced by von Neumann. It is also useful to refer to the Church–Turing thesis concerning the universality of computation, formulated by Turing in his 1936 paper, as well as to the recursive nesting of these different machines in the *bootstrap* mechanism.

The "red thread" that runs from the IT stack implementing a wide variety of languages to the user interfaces, whose "simplicity" is only apparent, in reality masks a formidable complexity that is reflected in the engineering costs of these interfaces, which amount to tens of millions of working hours.

In modelling, we most often start by describing what we want to model on a case-by-case basis, what we call a "use case", which is an extensional

---

22 This quote was found written on his blackboard at the time of his death in February 1988, as shown in a photo in the Caltech archives.

representation, understandable by users, before proceeding to a generalization in algorithmic form, when possible; we then obtain an intentional representation, more compact (think of the gains obtained when we went from the voluminous numerical tables, used by engineers before the era of calculators, to the algorithmic representations, available today in our smartphones; Laplace famously once quipped that the invention of logarithms has doubled the life span of astronomers).

In the theory of formal languages[23], there are equivalence theorems concerning the description of a language, either in the form of grammars (external language, for users) or in the form of automata (internal language, for engineering). From an "energetic" point of view, a text in grammatical form is far more "economical" than an equivalent text in the form of an automaton expressed in the language of the machine in a ratio that can be in the order of 1–100 (two orders of magnitude). Hence, the interest and economic importance of these equivalences in the engineering of complex systems, given the amplitude is such that one can go from the feasible to the infeasible[24]. If the architect does not "see" the abstractions underlying the formulation, he will miss the simplicity, in the sense of Feynman. One passes from the grammatical form, the most abstract, to the automated form, more concrete, by the techniques of compilation perfectly mastered today; one has the same notion with the duality functional language/imperative language. It is the implementation of Hilbert's "to simplify is to generalize".

In geometry, we know the importance of the choice of this or that system of coordinates (Cartesian, polar, affine, intrinsic, etc.), the reference frame, for the resolution of a geometrical problem, or even the choice of the most convenient geometry, Euclidean or not, as Poincaré pointed out. For example, the classical geometry of cylinders, with ruler and compass, is drastically simplified when formulated in terms of complex projective geometry, a preliminary step in the evolution toward Grothendieck's algebraic geometry.

With information engineering, what plays the role of referential, or coordinate system, are the underlying abstract machines and their expressive capacity (operational semantics, defined by the instruction set available on the machine); that is, their instruction sets, and this up to the transducers, which, ultimately, will realize the service requested from the system in the physical spatiotemporal reality of its real environment, integrating the corresponding hazards, including the errors and the "noise" of its environment. This we call a "meta-model".

---

23 See the reference work by Hopcroft, J. and Ullman, J. (1969). *Formal Languages and their Relation to Automata*, Addison-Wesley, Reading, MA.

24 In system engineering, it is the notion of affordability, that is to say the best that can be obtained for a given investment; i.e., the first E of PESTEL.

5.6. Computational volume: Remarks on the first quantification of complexity

The initial reflections of von Neumann on the quantity of computations necessary to solve a problem would lead, some 30 years later, to a much more detailed concept of "algorithmic complexity", a fundamental notion at the heart of the study on the performance of algorithms, and to the idea of the "logical depth of a computation" proposed by the physicist Charles Bennett[25], IBM Fellow. On these fundamental notions will be grafted other important problems, such as the reversibility of calculations, the loss of information, interface hierarchies, etc., notions that can be seen at work in structures such as interface stacks and API hierarchies.

In the last chapter of his book *The Game of Chance and Complexity*, Kourilsky concludes with a paragraph entitled "From Classical Biology to a 'Biology of Quanta'" in reference to the mutation that classical mechanics underwent when it passed from a continuous formulation, that of Lagrange and Hamilton, or even Einstein, to quantum mechanics, where physical systems evolve in a discrete way, or at least the perception that we have of them. The nature of these "biological quanta" remains mysterious[26], but it seems certain that they are discrete, because in the end, they are energetic exchanges that require a certain amount of time, such as the setting up of an appropriate immune response at the cellular level in order to make the "right" antibodies. Let us recall that in the quantum world, the order of operations is significant, as it is in computerized systems.

In a real computer, the operation $(a - b + c)$ is not equivalent to $(a + c - b)$! One must take into account the temporality and sequentiality of the operations which, moreover, are neither associative nor commutative, as the operation $(a + c - b)$ can trigger an *overflow*.

This is obviously a fundamental point to be clarified in a future work on the comparative complexity of biological systems and artificial systems that are now massively computerized. Let us simply note that the references in the biological literature to notions from theoretical computer science, such as "Bennett's depth", modularity or atomic processes, are sufficiently numerous to merit our attention (although we cannot, at this stage, judge their relevance to biology).

---

25 Bennett, C.H. (1988). Logical depth and physical complexity. In *The Universal Turing Machine: A Half-Century Survey*, Herken, R. (ed.). Oxford University Press, Oxford; see also Delahaye, J.-P. (2009). *Complexité aléatoire et complexité organisée*. Quae, Versailles.

26 Like the construction of a protein and the quality control that seems to be associated with it; see Kurilsky's work.

As far as computer science is concerned, we are indeed, from the outset, in a quantized world, since a computational "step", such as an addition, is a change in the state of the memory, following a "computational" process that, from two inputs (the operands), will compute the output result of the process, that is, $\{a,\ b\}$ — + →$\{r\}$, the process denoted plus indivisible (it is thus a quantization, in the strict sense, imposed by the organic structure of the machine). More generally, for any transformation, with or without memory, performed by the machine, we will have $\{e_1,\ e_2,\ e_n\}$ —— Operation→ $\{s_1,\ s_2,\ s_m\}$, a generalized instruction (see the generic CRUDE operations, above, which correspond to the basic instructions of a TM, notwithstanding the availability of resources necessary for the process.

Every operation therefore has an energy cost in terms of computing time and space, a cost that can vary from one operation to another, depending on the complexity of the operation and the specificities of the environment.. The arithmetic operations $\{$- +$\}$ are the least costly in terms of the number of *transistors*, then $\{x\}$, and finally division $\{\div\}$, among the several hundred that a last-generation processor has. Taking into account the statistical frequency of operations for a given type of application, we can define an average cost per operation, which will make it possible to calculate the average cost of a complete calculation process, which requires N elementary operations. In theoretical computer science, to simplify life, the reference machine is a TM and the calculation will be done in number of calculation steps by the TM, which makes it possible to establish calculation formulas such as the number of steps required to sort a list of n elements (for example, n numbers), traditionally written as $O[(n\ \mathrm{x log(n))}]$ to signify a logarithmic linear growth, to within one multiplicative and/or additive constant (this is the log of a factorial). The establishment of this kind of formula is an important chapter of theoretical computer science called *algorithmic complexity*.

Let us recall here that a TM is a mathematical artefact, necessary and sufficient for Turing's demonstration of the non-decidability of the decision problem as formulated by Hilbert; but in no case is it a computer model; we will have to wait for von Neumann's *logical model*. For Wittgenstein, the TM is a "human who calculates"!

5.6.1. *Quantifying interactions and functional dependencies*

In practice, if one wants to determine the true energy cost, TM is insufficient because other phenomena must be taken into account. One must reason within the framework of the von Neumann architecture and bring out the memory hierarchy, because in a real computer, there is always a memory hierarchy and inputs/outputs via the machine's "ports", which are external channels necessary for interactions with the environment. In other words, one must take into account the organization of the system; it is understood that this architectural organization – it is the information that has a cost of elaboration – has a determining and "dimensioning" influence on energy consumption.

As stated famously in Moore's Law, the share of performance that goes to the architecture of the machine is increasingly significant. Especially if we wish to make comparisons with the chemical universe of biology, we must distinguish the hierarchy of resources in decreasing "*energy*" costs:

– the internal memory (Mi) directly attached to the computing organ is the fastest, with associative memory devices (typically what we call "registers"); it is also the most expensive and difficult to build;

– the memory shared between different computing organs, often called *cache memory* (Mc), which will have a slightly worse access time than the internal memory, but which remains internal to the chip, in the case of a multicore – recall here that a shared memory of 1 megabyte (see above) can take a number of configurations of about $4.27 \times 10^{2.525.222}$, which is an inaccessible number impossible to visit;

– the machine's working memory (M_t or M_w) (*working storage* in the jargon), commonly called RAM, whose size today is in the gigabytes;

– the external memories that will be accessed via the ports of the machine: inputs/outputs, either toward storage disks associated with the machine, M_{eD}, or toward other storage devices accessed via Internet-type networks, M_{eR} or M_{eN} (*network*); that is the slowest, but the least expensive of all, tera-sized, or even as large as petabytes.

Between the internal memory, transistorized, and an external memory, on magnetic media, the performance differential in interaction *time*, based on the machine's clock, can be in the order of a million! It is therefore, in practice, unrealistic to neglect it, especially if we are aiming for a useful model.

The cost of a computational process – it is a "vector" quantity – will thus be characterized by a quintuplet (M_i, M_c, M_t, M_{eD} and M_{eR}), which will be different depending on whether one is:

– either in the initialization mode of the process, at the beginning the memories (M_i, M_c and M_t) are empty, and it is necessary to go to seek information in the external memories, which is thus slow, even very slow;

– or in a steady state, where ideally all the information would be in the M_i and/or M_t memories.

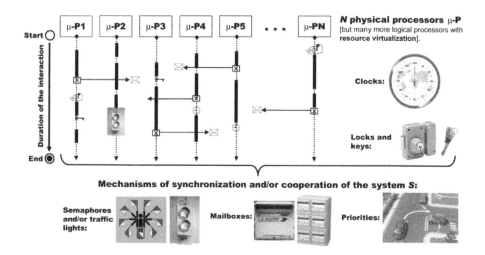

Figure 5.2. *An illustration of control energy used to manage module/thread cooperation*

All modern computers have more or less clever algorithms to manage the aging and/or the refreshment of the cache memories M_c (see, in particular, the algorithms implemented by search engines).

When the information is held by a human operator, it is necessary to access this operator in a secure way; it is the slowest operation that will be counted in seconds, in the best of cases, taking into account the human ergonomics, as is the case in control rooms.

In real systems, there is a last possibility of access, which would be a message sent by system S1 to system S2 asking it to perform, on its behalf, a job implementing a specific computational process in S2. In this case, the quintuplet S1 inherits the corresponding quintuplet in S2 in order to have the complete cost of the operation from the point of view of S1.

In an active quantified system (computer scientists also say "living"), today, the majority of computerized systems are, and this will make sense from an energy perspective:

– the number of quantified elementary entities, generally called *sequential processes* or *threads* (in UNIX jargon, these are processes qualified as "light") or simply "transactions" in computerized systems, and the frequency of solicitation for each of them (the average activation cost is therefore a weighted sum, i.e., a barycenter);

– the number of interactions to be carried out so that the elementary processes co-operate with each other in view of the goal to be reached (this is a control "energy" necessary for the organization of parallelism), so as to respond in an optimal way to the queries posed by the users (humans and/or other systems) of the active system (see the paragraph on *dynamic complexity* above).

One of the founding texts of parallelism is entitled *Cooperating Sequential Processes*, written by Dijkstra in 1968 and available online.

The architecture of the system and the cost of developing this architecture, which also includes process of building the system (this is the IVVT process, in system engineering), are fundamental characteristics of the complexity of the system and different, on the one hand, from algorithmic complexity and, on the other hand, from textual complexity, even if all of these are linked and interact.

The architecture is a highly improbable structure and, therefore, its quantity of information, in Shannon's sense, is very high (or, to speak like Brillouin, the negentropy is high, which corresponds to an important injection of order/ information, costly in energetic terms, i.e., a programing effort calculated in working hours).

Let us note that the introduction of asynchronism, controlled/regulated via the IPC synchronization language, is a particularly elegant way to visit combinatorics which, without it, would have remained perfectly inaccessible because the corresponding synchronous automaton that necessarily exists – what Borel calls an "imagined" number – is not realizable with the available memory resources, because it is inaccessible. The introduction of parallelism, from the first machines, is therefore an enormous simplification, well-understood by Turing and von Neuman[27], in terms of combinatorics, but a simplification that has a cost and limitations, because the properties expected by construction must be demonstrated (see the *model checking* approach already mentioned), which brings us back to the problem of the inaccessible numbers generated by combinatorics.

Moreover, in the engineering of real artificial systems, it is imperative to take into account errors, the severity and cost of repair of which generally follow a power law. In cases of anomalies noticed by the users/operators, it is necessary to be able to repair them, which poses the problem of the reversibility of the transformations carried out in order to go back to the cause of the anomaly and to the fact generation. In addition to the nominal system, a repair system is needed that takes into account

---

27 Obvious as far as von Neumann is concerned; for Turing, we think, in particular, of his article "Systems of logic based on ordinals" (1938), which he wrote while at Princeton, where he demonstrated a truly amazing anticipation of future problems.

the internal and external problems that can cause anomalies (error rates, environmental "noise", user behavior, etc.) and memorizes the intermediate states that are necessary and sufficient to guarantee reversibility to a coherent nominal state (the C in ACID transactions)[28], which always has an energy cost (the Ss system below); the temporality of the operations is structuring.

To illustrate this in detail, we can use the metaphor of the computing hierarchy (in Sifakis' terminology, a new structure derived from ICT) (see Printz (2012)).

What this specific ICT "information physics" highlights is the implementation of two types of "energies" that can perform any computational process in the real world:

– an energy that corresponds to the computational "step" (denoted here as E_C), well-understood by the algorithmic complexity (Bennett depth) $\rightarrow$ we count the elementary "atomic" operations and their costs;

– a transversal energy (denoted here as E_T) that corresponds to the cooperation/synchronization of the processes when the system is "alive", which enters into the framework of dynamic complexity $\rightarrow$ we count the transversal interactions inside/outside and the associated costs to which latency times must be integrated, as these can affect the survival of the system.

Therefore, there are two axes along which to pursue the mechanical metaphor:

– kinetic energy in the classical sense of the term, that is, $\frac{1}{2}m \times v^2$ 1, which everyone knows;

– the integral of the momentum and the angular momentum, which involves the rotation around an axis or a point, although in the world of trajectories, one must consider the direction (tangential energy to the trajectory) and the transversal energies corresponding to the curvature and the torsion of the trajectories, which are second-order energies (it is a tensorial quantity, resulting from the combination of several linear operations) in the 3D geometric space. The temporality of the operations is structuring.

Theoretical computer science, to simplify, does not take into account the transverse energy, since it reasons with infinite resources, as in the theory of perfect gases. In nature, everything is quantified, everything is dissipative, and one must of course take into account the exchanges with the outside world. A "closed" system is always a theoretical abstraction; it is a useful but insufficient limit case in the engineering of real systems; in real life, systems are open and dissipative.

---

28 See https://en.wikipedia.org/wiki/ACID.

In order to build robust/resilient systems, one must therefore consider the following complexities.

The nominal system (denoted here as S_N), given hazards and "noise", must be monitored by an ad hoc surveillance system (denoted here as S_S), and each implements both computational energy and transversal energy, as shown in Figure 5.3.

The complete complexity, from the user's point of view, is the sum of the two, it being understood that it is a feedback loop in the cybernetic sense of the term. The whole must be architected in such a way that it remains humanly controllable, that is, ideally, linear, if only through interposed systems ("rigorous" engineering in the sense given to this term by Sifakis, already mentioned), and economically optimal, taking into account the hazards (this is the *how good is good enough* in the PAC logic of Valiant mentioned above).

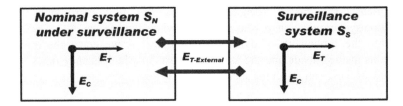

Figure 5.3. *Architecture of a robust/resilient system*

What is obvious from this type of analysis is that, if we only look at the transversal aspects (in systems engineering, this is what we call the traceability matrices of the 2×2 interactions, i.e., the functional dependencies, known as *N2* matrices), forgetting the hierarchy of interfaces as it appears in the IT stack, we obtain a perfectly incomprehensible *N2* matrix, which is also enormous because it adds up, without distinguishing them, all the levels of abstraction of the architecture in an undifferentiated mass, where all structure has disappeared. All the information concerning the structure is lost; reversibility is made impossible, except if one has preserved the traces of the transformations, which has a cost. The slightest failure will be fatal. As in living systems, where "to live, one must first survive", artificial systems can only guarantee the maintenance of their service contract if the safety of their operation has been finely analyzed (i.e., risk analysis or engineering) and integrated into the architecture, and for any situation that is deemed abnormal, there must be an applicable countermeasure. Robustness, as we say, is therefore a primary datum of the architecture, a constructed property and a fundamental aspect of complexity that brings out its relative and conventional side according to the mission (capability/PAC logic, a notion that is now classic in systems engineering, such as in C4ISTAR defense and security systems).

Systems engineering, formerly known as complex systems, whose founding act was the SAGE system developed by MIT in the 1950s, is part of an evolutionary perspective for which the INCOSE[29] document *A World in Motion: System Engineering Vision 2025* gives a good overview and sets the challenges.

5.7. References

Borel, E. (1911). *Le Hasard*. Librairie Félix Alcan, Paris.

Printz, J. (2012). *Architecture logicielle*. Dunod, Paris.

Printz, J. (2018). *Survivrons-nous à la technologie ?* Les acteurs du savoir, Paris.

Printz, J. (2020). *System Architecture and Complexity*. ISTE Ltd, London, and John Wiley & Sons, New York.

Printz, J. (2023). *Organization and Pedagogy of Complexity: Systemic Case Studies and Prospects*. ISTE Ltd, London, and Wiley, New York.

---

29 Available at: http://www.incose.org/AboutSE/sevision.

List of Authors

Thomas ANGLADE
ALGORITHMI
Paris
France

Thierry BERTHIER
CREC
École de Saint-Cyr
Guer
and
University of Limoges
France

Jean-Pierre BRIFFAUT
Institut F.R. Bull
Les Clayes-sous-Bois
France

Céline CHERICI
CHSSC
University of Picardy Jules Verne
Amiens
France

Jean-Paul DELAHAYE
CRISTAL
CNRS
University of Lille
France

Xosé M. FERNÁNDEZ
IQVIA Cancer Research
Cambridge
United Kingdom

Philippe KOURILSKY
Académie des Sciences
Paris
France

Jacques PRINTZ
Laboratoire ETHICS
Lille Catholic University
France

Index

Other titles from

in

Systems and Industrial Engineering – Robotics

2023

2022

BOURRIÈRES Jean-Paul, PINÈDE Nathalie, TRAORÉ Mamadou Kaba, ZACHAREWICZ Grégory
From Logistic Networks to Social Networks: Similarities, Specificities, Modeling, Evaluation

DEMOLY Frédéric, ANDRÉ Jean-Claude
4D Printing 1: Between Disruptive Research and Industrial Applications
4D Printing 2: Between Science and Technology

HAJJI Rafika, JARAR OULIDI Hassane
Building Information Modeling for a Smart and Sustainable Urban Space

KROB Daniel
Model-based Systems Architecting: Using CESAM to Architect Complex Systems
(Systems of Systems Complexity Set – Volume 3)

LOUIS Gilles
Dynamics of Aircraft Flight

2020

BRON Jean-Yves
System Requirements Engineering

KRYSINSKI TOMASZ, MALBURET FRANÇOIS
Energy and Motorization in the Automotive and Aeronautics Industries

PRINTZ Jacques
System Architecture and Complexity: Contribution of Systems of Systems to Systems Thinking

2019

ANDRÉ Jean-Claude
Industry 4.0: Paradoxes and Conflicts

BENSALAH Mounir, ELOUADI Abdelmajid, MHARZI Hassan
Railway Information Modeling RIM: The Track to Rail Modernization

BLUA Philippe, YALAOU Farouk, AMODEO Lionel, DE BLOCK Michaël,
LAPLANCHE David
Hospital Logistics and e-Management: Digital Transition and Revolution

BRIFFAUT Jean-Pierre
*From Complexity in the Natural Sciences to Complexity in Operations
Management Systems
(Systems of Systems Complexity Set – Volume 1)*

BUDINGER Marc, HAZYUK Ion, COÏC Clément
Multi-Physics Modeling of Technological Systems

FLAUS Jean-Marie
Cybersecurity of Industrial Systems

JAULIN Luc
Mobile Robotics – Second Edition Revised and Updated

KUMAR Kaushik, DAVIM Paulo J.
Optimization for Engineering Problems

TRIGEASSOU Jean-Claude, MAAMRI Nezha
*Analysis, Modeling and Stability of Fractional Order Differential Systems 1:
The Infinite State Approach
Analysis, Modeling and Stability of Fractional Order Differential Systems 2:
The Infinite State Approach*

VANDERHAEGEN Frédéric, MAAOUI Choubeila, SALLAK Mohamed,
BERDJAG Denis
Automation Challenges of Socio-technical Systems

2018

BERRAH Lamia, CLIVILLÉ Vincent, FOULLOY Laurent
*Industrial Objectives and Industrial Performance: Concepts and Fuzzy
Handling*

GONZALEZ-FELIU Jesus
Sustainable Urban Logistics: Planning and Evaluation

2017

ANDRÉ Jean-Claude
From Additive Manufacturing to 3D/4D Printing 1: From Concepts to Achievements
From Additive Manufacturing to 3D/4D Printing 2: Current Techniques, Improvements and their Limitations
From Additive Manufacturing to 3D/4D Printing 3: Breakthrough Innovations: Programmable Material, 4D Printing and Bio-printing

ARCHIMÈDE Bernard, VALLESPIR Bruno
Enterprise Interoperability: INTEROP-PGSO Vision

CAMMAN Christelle, FIORE Claude, LIVOLSI Laurent, QUERRO Pascal
Supply Chain Management and Business Performance: The VASC Model

FEYEL Philippe
Robust Control, Optimization with Metaheuristics

MARÉ Jean-Charles
Aerospace Actuators 2: Signal-by-Wire and Power-by-Wire

POPESCU Dumitru, AMIRA Gharbi, STEFANOIU Dan, BORNE Pierre
Process Control Design for Industrial Applications

RÉVEILLAC Jean-Michel
Modeling and Simulation of Logistics Flows 1: Theory and Fundamentals
Modeling and Simulation of Logistics Flows 2: Dashboards, Traffic Planning and Management
Modeling and Simulation of Logistics Flows 3: Discrete and Continuous Flows in 2D/3D

2016

ANDRÉ Michel, SAMARAS Zissis
Energy and Environment
(Research for Innovative Transports Set – Volume 1)

AUBRY Jean-François, BRINZEI Nicolae, MAZOUNI Mohammed-Habib
Systems Dependability Assessment: Benefits of Petri Net Models
(Systems Dependability Assessment Set – Volume 1)

BLANQUART Corinne, CLAUSEN Uwe, JACOB Bernard
Towards Innovative Freight and Logistics
(Research for Innovative Transports Set – Volume 2)

COHEN Simon, YANNIS George
Traffic Management
(Research for Innovative Transports Set – Volume 3)

MARÉ Jean-Charles
Aerospace Actuators 1: Needs, Reliability and Hydraulic Power Solutions

REZG Nidhal, HAJEJ Zied, BOSCHIAN-CAMPANER Valerio
Production and Maintenance Optimization Problems: Logistic Constraints
and Leasing Warranty Services

TORRENTI Jean-Michel, LA TORRE Francesca
Materials and Infrastructures 1
(Research for Innovative Transports Set – Volume 5A)
Materials and Infrastructures 2
(Research for Innovative Transports Set – Volume 5B)

WEBER Philippe, SIMON Christophe
Benefits of Bayesian Network Models
(Systems Dependability Assessment Set – Volume 2)

YANNIS George, COHEN Simon
Traffic Safety
(Research for Innovative Transports Set – Volume 4)

2015

AUBRY Jean-François, BRINZEI Nicolae
Systems Dependability Assessment: Modeling with Graphs and Finite State
Automata

BOULANGER Jean-Louis
CENELEC 50128 and IEC 62279 Standards

STEFANOIU Dan, BORNE Pierre, POPESCU Dumitru, FILIP Florin Gh.,
EL KAMEL Abdelkader
*Optimization in Engineering Sciences: Metaheuristics, Stochastic Methods
and Decision Support*

2013

ALAZARD Daniel
Reverse Engineering in Control Design

ARIOUI Hichem, NEHAOUA Lamri
Driving Simulation

CHADLI Mohammed, COPPIER Hervé
Command-control for Real-time Systems

DAAFOUZ Jamal, TARBOURIECH Sophie, SIGALOTTI Mario
Hybrid Systems with Constraints

FEYEL Philippe
Loop-shaping Robust Control

FLAUS Jean-Marie
Risk Analysis: Socio-technical and Industrial Systems

FRIBOURG Laurent, SOULAT Romain
*Control of Switching Systems by Invariance Analysis: Application to Power
Electronics*

GROSSARD Mathieu, REGNIER Stéphane, CHAILLET Nicolas
Flexible Robotics: Applications to Multiscale Manipulations

GRUNN Emmanuel, PHAM Anh Tuan
Modeling of Complex Systems: Application to Aeronautical Dynamics

HABIB Maki K., DAVIM J. Paulo
*Interdisciplinary Mechatronics: Engineering Science and Research
Development*

HAMMADI Slim, KSOURI Mekki
Multimodal Transport Systems

DOUMIATI Moustapha, CHARARA Ali, VICTORINO Alessandro, LECHNER Daniel
Vehicle Dynamics Estimation using Kalman Filtering: Experimental Validation

GUERRERO José A, LOZANO Rogelio
Flight Formation Control

HAMMADI Slim, KSOURI Mekki
Advanced Mobility and Transport Engineering

MAILLARD Pierre
Competitive Quality Strategies

MATTA Nada, VANDENBOOMGAERDE Yves, ARLAT Jean
Supervision and Safety of Complex Systems

POLER Raul *et al.*
Intelligent Non-hierarchical Manufacturing Networks

TROCCAZ Jocelyne
Medical Robotics

YALAOUI Alice, CHEHADE Hicham, YALAOUI Farouk, AMODEO Lionel
Optimization of Logistics

ZELM Martin *et al.*
Enterprise Interoperability –I-EASA12 Proceedings

2011

CANTOT Pascal, LUZEAUX Dominique
Simulation and Modeling of Systems of Systems

DAVIM J. Paulo
Mechatronics

DAVIM J. Paulo
Wood Machining

GROUS Ammar
Applied Metrology for Manufacturing Engineering

KOLSKI Christophe
Human–Computer Interactions in Transport

LUZEAUX Dominique, RUAULT Jean-René, WIPPLER Jean-Luc
Complex Systems and Systems of Systems Engineering

ZELM Martin, *et al.*
Enterprise Interoperability: IWEI2011 Proceedings

2010

BOTTA-GENOULAZ Valérie, CAMPAGNE Jean-Pierre, LLERENA Daniel,
PELLEGRIN Claude
Supply Chain Performance: Collaboration, Alignment and Coordination

BOURLÈS Henri, GODFREY K.C. Kwan
Linear Systems

BOURRIÈRES Jean-Paul
Proceedings of CEISIE'09

CHAILLET Nicolas, REGNIER Stéphane
Microrobotics for Micromanipulation

DAVIM J. Paulo
Sustainable Manufacturing

GIORDANO Max, MATHIEU Luc, VILLENEUVE François
Product Life-Cycle Management: Geometric Variations

LOZANO Rogelio
Unmanned Aerial Vehicles: Embedded Control

LUZEAUX Dominique, RUAULT Jean-René
Systems of Systems

VILLENEUVE François, MATHIEU Luc
Geometric Tolerancing of Products

2009

DIAZ Michel
Petri Nets: Fundamental Models, Verification and Applications

OZEL Tugrul, DAVIM J. Paulo
Intelligent Machining

PITRAT Jacques
Artificial Beings

2008

ARTIGUES Christian, DEMASSEY Sophie, NERON Emmanuel
Resources–Constrained Project Scheduling

BILLAUT Jean-Charles, MOUKRIM Aziz, SANLAVILLE Eric
Flexibility and Robustness in Scheduling

DOCHAIN Denis
Bioprocess Control

LOPEZ Pierre, ROUBELLAT François
Production Scheduling

THIERRY Caroline, THOMAS André, BEL Gérard
Supply Chain Simulation and Management

2007

DE LARMINAT Philippe
Analysis and Control of Linear Systems

DOMBRE Etienne, KHALIL Wisama
Robot Manipulators

LAMNABHI Françoise *et al.*
Taming Heterogeneity and Complexity of Embedded Control

LIMNIOS Nikolaos
Fault Trees

2006

FRENCH COLLEGE OF METROLOGY
Metrology in Industry

NAJIM Kaddour
Control of Continuous Linear Systems

Printed and bound by CPI Group (UK) Ltd, Croydon, CR0 4YY

16/04/2025

14658458-0001